REVITALISING US–RUSSIAN SECURITY COOPERATION
Practical Measures

RICHARD WEITZ

ADELPHI PAPER 377

FIRST PUBLISHED NOVEMBER 2005
BY **Routledge**
4 PARK SQUARE, MILTON PARK, ABINGDON, OXON, OX14 4RN
FOR **The International Institute for Strategic Studies**
ARUNDEL HOUSE, 13–15 ARUNDEL STREET, TEMPLE PLACE, LONDON, WC2R 3DX
WWW.IISS.ORG

Simultaneously published in the USA and Canada
by **Routledge**
270 MADISON AVE., NEW YORK, NY 10016

Routledge is an imprint of the Taylor & Francis Group

© 2005 THE INTERNATIONAL INSTITUTE FOR STRATEGIC STUDIES

DIRECTOR John Chipman
EDITOR Tim Huxley
MANAGER FOR EDITORIAL SERVICES Ayse Abdullah
COPY EDITOR Matthew Foley
PRODUCTION Jesse Simon
COVER PHOTOGRAPH Getty Images News

TYPESET BY Techset Composition Ltd, Salisbury, Wiltshire
PRINTED AND BOUND IN GREAT BRITAIN BY Bell & Bain Ltd, Thornliebank, Glasgow

British Library Cataloguing in Publication Data
A catalogue record for this book is available from the British Library

Library of Congress Cataloguing in Publication Data

ISBN 0-415-39864-9
ISSN 0567-932X

CONTENTS

Glossary

ABM	Anti-Ballistic Missile
ALTBMD	Active Layered Theatre Ballistic Missile Defence
AMEC	Arctic Military Environmental Cooperation
BMD	ballistic-missile defence
BW	biological weapons
CFE	Conventional Forces in Europe Treaty
CIA	Central Intelligence Agency
CNAD	NATO Conference of National Armaments Directors
CONOPS	Concept of Operations
CPX	Command Post Exercise
CTBT	Comprehensive Test Ban Treaty
CTR	Cooperative Threat Reduction
CW	chemical weapons
CWC	Chemical Weapons Convention
CWDF	Chemical Weapons Disposition Facility
DOD	US Department of Defense
DOE	US Department of Energy
EAPC	NATO Euro-Atlantic Partnership Council
EMERCOM	Russian Ministry for Civil Defence, Emergencies and the Elimination of the Consequences of Natural Disasters
EU	European Union
FBI	US Federal Bureau of Investigation
FEMA	US Federal Emergency Management Agency
FMSF	Fissile Material Storage Facility
FSB	Russian Federal Security Service
FSTEK	Russian Federal Technical and Export Control Service
FSU	Former Soviet Union
G-8	Group of Eight industrial states
GAO	US Government Accountability Office; formerly General Accounting Office
GDP	gross domestic product
GIPP	Global Initiatives for Proliferation Prevention
GTRI	Global Threat Reduction Initiative
GWOT	Global War on Terrorism
HEU	highly-enriched uranium
IAEA	International Atomic Energy Agency
IED	improvised explosive device
IFOR	NATO Implementation Force in Bosnia-Herzegovina

IMET	International Military Education and Training Program
INF	Intermediate-Range Nuclear Forces
ISTC	International Science and Technology Centre
KFOR	NATO Kosovo Force
LEU	low-enriched uranium
MDA	US Missile Defense Agency
MNEPR	Multilateral Nuclear Environmental Programme in the Russian Federation
MOX	Mixed-oxide (fuel)
MTCR	Missile Technology Control Regime
NACC	North Atlantic Cooperation Council
NASA	National Aeronautics and Space Agency
NATO	North Atlantic Treaty Organisation
NCI	Nuclear Cities Initiative
NMD	national missile defence
NNSA	National Nuclear Security Administration
NPR	Nuclear Posture Review
NPT	Nuclear Non-proliferation Treaty
NRC	NATO–Russia Council
OSD	Office of the Secretary of Defense
OSCE	Organisation for Security and Cooperation in Europe
PfP	NATO Partnership for Peace
PJC	NATO Permanent Joint Council
PNI	Presidential Nuclear Initiatives
PSI	Proliferation Security Initiative
RAMOS	Russian–American Observation Satellite
RERTR	Reduced Enrichment for Research and Test Reactors
RTG	radioisotope thermoelectric generator
SACEUR	NATO Supreme Allied Commander, Europe
SFOR	NATO Stabilisation Force in Bosnia-Herzegovina
SHAPE	Supreme Headquarters Allied Powers Europe
SIG	Senior Interagency Group
SOFA	Status of Forces Agreement
SORT	Strategic Offensive Reductions Treaty
SSBN	ballistic-missile submarine
STANAG	NATO Standardisation Agreement
TMD	theatre-missile defence
TMD AHWG	NATO Ad-Hoc Working Group on Theatre-Missile Defence
TNW	tactical nuclear weapons
WMD	weapons of mass destruction
WSSX	Warhead Safety and Security Exchange Agreement

With Russia, we are already building a new strategic relationship based on a central reality of the twenty-first century: the United States and Russia are no longer strategic adversaries ... At the same time, we are realistic about the differences that still divide us from Russia and about the time and effort it will take to build an enduring strategic partnership.

National Security Strategy of the United States
(Washington DC: The White House, 2002), chapter 8.

It has been stated they aren't an enemy, but they aren't allies either, that's for sure.

Russian Defence Minister Sergei Ivanov,
referring to the United States in an interview with
Moskovskiy Komsomolets, 28 October 2003.

Introduction

A decade and a half after the end of the Cold War, Russia and the United States are still the world's two most important states for many vital security issues. They possess the world's largest nuclear arsenals, they are involved in the world's principal regional conflicts and they play leading roles in opposing international terrorism and the proliferation of weapons of mass destruction (WMD). In most of Eurasia, their interests overlap, even if they do not always coincide.

Enhanced cooperation between these two key countries could help avert and resolve conflicts, counter terrorist threats and curb the spread of dangerous technologies. Despite significant bilateral tensions and other impediments, opportunities exist for improving security cooperation between Russia and the United States. Since short-term results in the areas of formal arms control or ballistic-missile defence are unlikely, the two governments should focus on improving and expanding their joint threat-reduction and non-proliferation programmes, enhancing their military dialogue regarding Central Asia, strengthening the links between their defence industries and deepening their anti-terrorist cooperation, both bilaterally and through NATO. Making greater use of market incentives, expanding reciprocity and limiting disputes over Russia's nuclear cooperation with Iran would all facilitate progress.

Constraints and opportunities

Fundamental domestic and international political conditions warrant modest expectations regarding the scale and scope of likely improvements in bilateral security ties. On the American side, compromises with Russia have typically been difficult to promote domestically, and Russian President Vladimir Putin's authoritarian traits have made sustaining support for security cooperation harder still. In the latest edition of its global survey, the NGO Freedom House characterises Russia as 'Not Free'.[1] The Organisation for Security and Cooperation in Europe (OSCE) states that the presidential elections in March 2004 'did not adequately reflect principles necessary for a healthy democratic election'.[2] Prominent Americans such as Republican Senator John McCain and Democratic Senator Joe Lieberman fear that Putin has exploited the Global War on Terrorism (GWOT) to expand his powers domestically and crack down in Chechnya without the risk of a major international backlash. Members of the US Congress have called for Russia's expulsion from the Group of Eight (G-8) industrial states because of its undemocratic tendencies.[3] Despite a general desire to achieve better security ties with Moscow, the Bush administration's new-found focus on promoting democracy abroad has made the problem of Putin increasingly vexing.[4]

On the Russian side, several issues impede bilateral security cooperation. The armed forces are more isolated from foreign influence than most Russian institutions. Strenuous efforts by senior Russian and US policymakers have often been necessary to secure the Russian military's involvement in international projects. The lack of security transparency within Russia further complicates efforts at engagement.[5] Russia's armed forces still conduct large-scale exercises with scenarios implying a nuclear war with a US-led coalition.[6] For their part, Russians fear that Americans have been using the GWOT to expand US power and influence around Russia's periphery. They also complain that successive US presidents have repeatedly failed to fulfill pledges to remove the humiliating, if largely symbolic, Jackson–Vanik amendment of 1974, which restricts bilateral trade.[7] Finally, they argue that Western governments have shown insufficient flexibility in negotiating Russia's entry into the World Trade Organisation (WTO), and have failed to ensure that aid intended for Russia is actually spent there. Meanwhile, NATO's ability to affect Russian policies is limited by the widespread understanding that Russia will not soon, and may never, become a full alliance member. As a result, Russians tend to view each wave of NATO expansion, and each successful alliance military operation, as a potential threat. Differences persist regarding the Conventional Forces in Europe (CFE) Treaty, the alliance's expansion and basing plans and US nuclear weapons in Europe.

At the same time, the experience of the past decade has shown that mutual interests regularly drive Russians and Americans to work together to overcome these impediments. Both governments must cooperate to achieve their goals regarding nuclear security, international terrorism and the proliferation of WMD. For the foreseeable future, Russia and the United States will possess the world's largest nuclear arsenals, giving them decisive importance in international affairs.[8] They alone have the power to destroy any other country, including each other. The Russian and US militaries may no longer constantly target each other, but they clearly take the other country into account when structuring their own forces. Keith Payne, who helped to write the US Nuclear Posture Review (NPR) of 2001, denies that concerns about Russia determined the NPR's recommended force level of 1,700–2,200 operationally deployed warheads.[9] But a leaked secret paragraph in the NPR states that: 'Russia's nuclear forces and programs, nevertheless, remain a concern. Russia faces many strategic problems around its periphery and its future course cannot be charted with certainty. US planning must take this into account. In the event that US relations with Russia significantly worsen in the future, the US may need to revise its nuclear force levels and posture.'[10]

US Department of Defense (DOD) officials have made evident that hedging against adverse changes in Russia and countering a possible future 'hostile peer competitor' prompted them to recommend retaining a large strategic arsenal.[11] Those involved in formulating nuclear policy during the Clinton administration cited concerns about a potential Russian threat even more openly.[12]

In Russia, analysts inside and outside the government continue to debate the importance of maintaining a robust nuclear deterrent.[13] Some Russians argue that they should concentrate on developing conventional forces, but the majority see nuclear weapons as essential for preserving Russia's status, especially since the other nuclear-weapon states evince little inclination to relinquish their own arsenals. Substantially enhancing Russia's conventional forces would entail much greater expenditure than maintaining a credible nuclear deterrent. Concerns about a potential future Chinese challenge to Russian interests in Asia reinforce the perceived need to retain robust nuclear forces.[14] In recent years, the government has adopted long-term plans to revitalise Russia's strategic nuclear weapons.[15]

Besides their nuclear arsenals, Russia and the United States possess more WMD-related material than any other country. Preventing the transfer of their weapons-grade uranium and plutonium, dangerous biological pathogens and chemical agents and ballistic missiles and other weapon systems to terrorists or countries of proliferation concern will remain an international priority for years. The Joint Statement on Nuclear Security Cooperation, issued by US President George W. Bush and Putin at their summit in Bratislava, Slovakia, in February 2005, rightly affirmed that both governments 'bear a special responsibility for the security of nuclear weapons and fissile material'. They also have special roles to play in safeguarding other WMD-related items, both within their countries and globally. In May 2004, Russia reaffirmed its support for US-led efforts to curb illicit trafficking in WMD, ballistic missiles and related items by joining the Proliferation Security Initiative. Russia's entry into this voluntary multilateral arrangement builds on a long history of collaboration between Moscow and Washington on non-proliferation issues, albeit one punctuated by acute differences in perception and priorities.[16] For countering WMD proliferation, Russia and the United States are the two indispensable nations.

Russian representatives have made clear that they prefer to work directly with their US counterparts because of Washington's unique status in relation to many global security issues. Besides their shared interest in curbing WMD proliferation, Russians see Americans as their leading ally in the GWOT. Russia and the United States have suffered more casualties from radical Islamist-inspired terrorism during the last five years than any other country.[17] They also belong to a select group of states that have experienced attacks within their territory involving chemical, biological or radiological agents: the mass mailing of anthrax in the United States in late 2001, and the radioactive cesium-137 left in Moscow's Izmailovsky Park by Chechen terrorists in November 1995.[18] More generally, Russians appreciate how America's worldwide interests lead it to engage Moscow on global issues – as opposed to the regionally focused dialogue Russian leaders normally conduct with their European and Asian counterparts.[19]

Casting aside earlier frictions, Putin was the first foreign leader to call Bush after the 11 September 2001 ('9/11') attacks to offer condolences and assistance. The Russian government subsequently supported the US-led *Operation Enduring Freedom* against the Taliban government and its al-Qaeda allies by sharing intelligence about Afghanistan with the United States, cancelling military exercises that might have distracted the US response, increasing military assistance to the anti-Taliban Northern Alliance and informing Central Asian governments that Russia would not object to Washington acquiring temporary military bases on their territories. In preparing for his November 2001 visit to the United States, Putin allegedly told a Russian parliamentarian that his relationship with Bush would soon resemble that between Winston Churchill and Franklin Roosevelt during the Second World War.[20]

Cooperation between Russian and US intelligence and law-enforcement agencies has helped to disrupt terrorist organisations, international criminal groups and other transnational security threats. The Mutual Legal Assistance Treaty of January 2002 has provided a basis for bilateral cooperation on identifying, seizing and freezing criminal or terrorist assets. Since then, the two governments have held numerous sessions of a joint working group on combating organised crime, including cyber-crime and illicit trafficking in people. At their summit in Moscow in May 2002, Bush and Putin expanded the mandate of the US–Russia Working Group on Afghanistan, and renamed it the US–Russia Working Group on Counter-terrorism. A breakthrough in bilateral law-enforcement cooperation occurred in August 2003, when the Russian Federal Security Service (FSB) and the US Federal Bureau of Investigation (FBI) cooperated to arrest a British national trying to smuggle Russian-made Man-Portable Air Defence Systems (MANPADS) into the United States, to be used to attack civilian airliners. The operation highlighted ten years of deepening cooperation between the FSB and the FBI.[21]

Security cooperation between Russia and the United States is essential for resolving several important regional conflicts. Russians have substantial influence in, and knowledge of, many regions and countries of strategic significance to the United States, including Central Asia, China, Iran and North Korea. Russian and US officials have collaborated to stabilise the post-Taliban government in Afghanistan, curb North Korea's nuclear ambitions and induce Syria to withdraw its military from Lebanon. Russia is a core member of both the Contact Group for the former Yugoslavia and the 'Quartet' seeking a Middle East peace. Washington and Moscow differ over the conflicts in Georgia and the Caucasus, and disagree profoundly on Iran's possible nuclear-weapons programme, but solving these disputes will require cooperative initiatives. Russia's membership of the G-8 (which it chairs in 2006), the UN Security Council (where it enjoys a right of veto) and other multilateral institutions gives its representatives influence over many important international questions.

The Russian economy has rebounded sharply since the financial crisis of August 1998. From 1998 to 2003, Russia's gross domestic product (GDP) expanded at an average annual rate of 6.5%, faster than any other G-8 member.[22] In 2004, GDP grew by 6.8% (to $613 billion), and real incomes increased by 10%.[23] Russia's trade surplus with the United States was $3.4bn in 2004, and Americans consistently rank among the leading investors in Russia. However, whereas Russians see the United States as a potential economic as well as defence partner, Americans remain primarily interested in Russia for security reasons. Despite recent growth, Russia's economy amounts to only one-tenth the size of American GDP, and approximates what the United States spends on national security alone. US economic engagement with Russia remains modest compared with many other countries, while the government's crackdown on prominent private entre-preneurs, its seemingly discriminatory regulatory policies and charges of corruption have reduced foreign investment and accelerated capital flight.[24] Collaborative development of high-technology defence products could create additional bilateral economic opportunities in civilian space launches, nuclear energy or homeland security. Russian–US cooperation in civilian outer space is already extensive, and includes both government projects and private-sector initiatives.[25]

The Bush administration's clear aversion to negotiating additional formal bilateral strategic arms-reduction or operational arms-control measures provides another reason to revisit the issue of Russian–US security coopera-tion. This reluctance partly reflects the major transformation that has already occurred in the defence relationship, manifested in the May 2002 Strategic Offensive Reductions Treaty (SORT; *Dogovor o sokrashshenii strategicheskix nastupatel'nix potentsalov* (SNP)). SORT provides for major reductions in both sides' nuclear arsenals, to between 1,700 and 2,200 'operationally deployed strategic warheads' by 31 December 2012. Observers complain about the treaty's lack of detailed verification procedures, the absence of a timetable for warhead reductions, its 90-day withdrawal clause and the fact that its limits take effect and expire on the same day.[26] Nevertheless, both govern-ments had previously signalled their intention to cut their forces unilaterally to such levels, and have continued to reduce them. According to Russian officials, between November 2000 and January 2005 Russia eliminated 1,740 nuclear warheads and 357 strategic delivery vehicles.[27]

Additional formal Russian–US arms-control accords are unlikely for the remainder of the Bush administration. In 2004, then National Security Advisor Condoleezza Rice stated that: 'We believe that [SORT] is a transi-tional measure to a day when arms control will play a very minor role in US–Russian relations, if a role at all'.[28] Washington has refused to ratify the Comprehensive Test Ban Treaty (CTBT), and opposes efforts by Russia and others to broaden restrictions on military activities in space.[29] Proposals for more operational arms control – such as lowering the readiness state of strategic forces, restricting the number of nuclear ballistic-missile subma-rines (SSBNs) on patrol and separating nuclear warheads from their means of delivery – have gained little support within the US administration.[30] Accordingly, this paper focuses on other areas and mechanisms for enhanc-ing Russian–American security relations.

Overview

The next chapter offers recommendations on how to improve bilateral threat-reduction programmes between the United States and Russia. These projects represent one of the most successful examples of peace-time security cooperation between major powers. They have advanced both parties' interests and, more generally, they have made the world a safer place. Yet they have also been plagued with problems. For example, intelligence concerns have led Russian officials to limit US access to WMD sites. These restrictions conflict with American laws that require on-site visits to verify the proper expenditure of US funds. Possible solutions to these access disputes include giving Russian representatives more oppor-tunities to see US WMD-related sites, hiring Russian firms or personnel to dismantle WMD stocks in the United States and supplying additional data concerning US-funded threat-reduction projects in Russia in return for more detailed information about Russia's WMD-related facilities and employees. Both sides could also relax their rules for granting visas to inspectors. With time, improvements in monitoring technology could reduce the need for American inspectors.

Exploiting market-based incentives might also improve the threat-reduction relationship. The relatively smooth implementation of the HEU–LEU Agreement of 1993 is largely due to the billions of dollars Russia accrues from the sale of its blended-down highly-enriched uranium for use as low-enriched uranium fuel in US nuclear reactors. The employment of Russian special-purpose equipment and subcontractors for US-funded threat-reduction programmes has lowered costs, developed local constitu-encies for projects and transferred valuable commercial skills to Russian firms. Market-based incentives are not optimal or even applicable in all cases; for example, non-market barriers and disincentives often work more effectively than financial considerations in discouraging black-market WMD-related entrepreneurship. In addition, programmes to redirect former biological-weapons scientists into non-defence work have encountered severe commercial barriers, suggesting that Russia and the United States might better use the expertise of these scientists to improve their defences against biological terrorism. While some elements of Russia's weapons complex can and should be diversified, other personnel and firms can best contribute to shared goals by continuing work on defence-related projects.

Both Russian and US officials appreciate the need to increase Russia's financial and other contributions to threat-reduction activities. Priority areas include bolstering biosecurity and accelerating plutonium disposi-tion. Increasing Russian support for threat-reduction initiatives could have several beneficial effects in addition to the immediate impact on improving the financial health of the programmes. First, it would affirm Russia's commitment to non-proliferation in general. Second, it would elevate Russia's status to that of a genuine partner in a common endeavour. Third, programme integration could improve if Russians' role in designing and implementing projects increased due to this enhanced status. Finally, greater Russian government support through enhanced financial involvement is essential for sustaining threat-reduction programmes in the long term. Restructuring the Russian and US threat-reduction bureaucracies could also strengthen the integration and implementation of these programmes. In particular, each government should designate a senior official to improve the development and application of threat-reduction projects. This individual should enjoy direct presidential access, influence funding decisions and have the authority to determine the roles, methods of interaction and procedures for resolving disputes and sharing information among the various agencies involved in threat-reduction projects. Reviving an institution like the high-level Gore–Chernomyrdin Commission could also enhance non-proliferation efforts. Finally, joint or parallel oversight of programmes by both national legislatures could reduce misperceptions and misunderstandings.

The Russian–US defence relationship has made notable progress in recent years. The Russian and American militaries have conducted several joint operations, and have discussed possible combined anti-terrorist operations. In Central Asia, however, their military contingents operate independently, with little direct communication. They should evaluate the utility of bilateral exercises and other joint activities. The Russian military's relative isolation from outside contacts and influence makes US attempts to engage the Russian defence community all the more essential. The armed forces will play a decisive role in shaping Russia's future domestic and foreign policies, and the Pentagon enjoys unique advantages in trying to affect its evolution. Military contacts and other forms of bilateral defence diplomacy are thus both necessary and possible.

Working together, the Russian and US defence research, development and acquisition communities could produce mutually beneficial military technologies and systems. Opportunities for effective collaboration appear especially promising in defences against harmful biological agents, improvised explosive devices, radiological dispersal devices and other forms of nuclear terrorism. Thanks to initiatives like the G-8 Global Partnership Against the Spread of Weapons and Materials of Mass Destruction and the US Global Threat Reduction Initiative, Russian and US non-proliferation experts are in a better position to collaborate on threat-reduction activities outside Russia. In countries such as Iran and North Korea, the local authorities might be more willing to cooperate with Russian personnel, who have long had a presence in these countries, than with Americans. Russia might also help to counter nuclear proliferation by repatriating and storing 'spent' (i.e., already used) nuclear fuel from these and other states, as they have alreasdy agreed to do for Iran.

Russian and US officials have repeatedly clashed over Russia's defence and dual-use exports. Washington has been especially worried about the transfer of Russian-origin WMD- and missile-related technologies to anti-American governments. Russian representatives dismiss such concerns as intended to curb unwelcome commercial competition. They have, however, strengthened their export-control procedures. The two governments also work through international institutions such as the UN to counter the spread of WMD and related technologies.

Russia's concerns about US and NATO plans to develop ballistic-missile defences (BMD) have declined, and Russian companies have sought to sell BMD technologies to Western governments, but substantial obstacles have prevented much progress. Russian and US officials disagree regarding the nature of the ballistic-missile threat in general, and Washington's intention to deploy BMD assets in Europe in particular. Russia and NATO have developed air- and missile-defence systems that employ different technical standards, command-and-control procedures and operational engagement doctrines, and have only recently undertaken initiatives to overcome these interoperability problems. Russian officials have made clear that investing in BMD-related technologies is not a defence priority.

Besides theatre-missile defence, the NATO–Russian relationship provides Moscow and Washington with a useful multilateral framework to supplement bilateral exchanges. The members of the NATO–Russian Council (NRC) have collectively assessed terrorist threats, and have shared intelligence and best security practices to counter them. They have also conducted numerous joint exercises to improve their defences against terrorist attacks, and to manage their consequences. The NATO–Russia Status of Forces Agreement of April 2005 will provide a more secure legal basis for joint military activities. NATO and Russian representatives are especially interested in deepening cooperation in peacekeeping and post-conflict stability operations. For example, they are drafting a Generic Concept for Joint NATO–Russian Peacekeeping Operations. Moscow's decision to form a separate brigade specially trained to operate with foreign militaries testifies to its interest in this area.

NATO and Russian defence planners have made some progress in overcoming their interoperability problems, particularly in the maritime realm. The NRC defence industry and technology working group and the NATO Research and Technology Organisation have provided mechanisms for defence-industry collaboration between Russian and Western companies and experts. Enhanced technical interoperability would facilitate Russian involvement in NATO military operations, and would expand export opportunities in Western defence markets. Thus far, however, such cooperation has been limited.

This mixed record reaffirms the need for modest expectations regarding prospects for short-term improvements in Russian–American security cooperation. Impediments to much deeper ties will remain, but there will also be opportunities for mutual gains from further defence collaboration. Russia and the United States will not in the near future become close allies, but they should be able to achieve better security ties given that, on most questions, their shared interests still outweigh the issues that divide them.

CHAPTER ONE

US–Russian threat-reduction programmes

In terms of dollars spent, the most important dimension of Russian–American security cooperation has been the wide variety of programmes aimed at securing Russia's WMD, WMD-related materials and facilities, and the people whose knowledge poses a serious proliferation threat. Altogether, the US Congress has allocated over $9bn to threat-reduction and non-proliferation programmes in the former Soviet Union (FSU).[1] Concerns about the ability of these newly independent states to fulfil their arms-control obligations and manage their WMD inheritance led Congress to enact the Nunn–Lugar programme in November 1991 to provide financial support for such efforts. In 1993, the DOD initiated the follow-on Cooperative Threat Reduction (CTR; *Programma sovmestnogo umen'sheniya ugrozi*) process. In 1997, other US government agencies took charge of their own projects.[2] Although the Departments of Defense, Energy and State finance most projects, additional US government agencies participate in niche areas (the Department of Agriculture, for instance, works with Russian biological-warfare experts).

Among their many achievements, these projects have helped to create the infrastructure to dismantle the FSU's strategic weapons. They have also enhanced the security and safety of these countries' WMD and WMD-related materials, impeded the trafficking of WMD-related items across their borders, redirected former Soviet military enterprises and scientists into civilian sectors and promoted defence contacts. Yet further progress is necessary. A US intelligence report of 2004, while acknowledging

Russia's improvements in 'upgrading its physical, procedural, and technical measures to secure its nuclear weapons against both external and internal threats', nevertheless expressed continued concern about certain 'risks' and 'vulnerabilities', and concluded that 'undetected smuggling' of weapons-usable nuclear material 'has occurred'.[3]

Mutual benefits

Joint threat-reduction programmes between Russia and the United States clearly advance the security interests of both countries. The 9/11 attacks and the growing fear that terrorists might use WMD have increased Russian and American interest in improving the security of their WMD, and related materials and knowledge (including that embodied in their scientists, engineers and other technical personnel). Given international unease about 'loose nukes', Russian representatives have used threat-reduction activities to dispel concerns about the vulnerability of their arsenals. In July 2004, Russian Defence Minister Sergei Ivanov said that he had invited foreign observers to monitor upcoming security exercises at Russian nuclear facilities 'to show that the existing myths about Russia's problems on this direction are really myths'.[4] Frequent interaction in this area helps to promote bilateral dialogue on non-proliferation, increases the attention paid to these issues in both countries and appears to deepen mutual trust between Russian and American military personnel, government officials, scientists and private contractors. The main US military commands employ CTR funds to finance exchanges with defence policymakers in Russia and other FSU states.[5] Through these programmes, the parties often obtain more information about each other's WMD-related capabilities and policies than they acquire through formal arms-control accords.[6]

Recurring criticisms

The benefits provided by these bilateral threat-reduction programmes have enabled them to survive major disputes in other areas. Nevertheless, representatives from both countries have expressed concerns. A perennial refrain is the limited financing of these programmes, given the magnitude of the challenge. In January 2001, a US Department of Energy (DOE) taskforce recommended that the US government triple spending on threat-reduction programmes in the FSU, to $3bn a year for eight to ten years.[7] In 2004, the National Commission on Terrorist Attacks Upon the United States ('the 9/11 Commission') argued that the CTR programme was 'now in need of expansion, improvement and resources'.[8] In September 2005, the leadership of the Democratic Party in the US House of Representatives called for tripling spending on CTR and other non-proliferation programmes.[9]

Despite these widely publicised recommendations, and notwithstanding the prominence the issue received during the 2004 US presidential elections and recurring American intelligence warnings about insecure Russian nuclear materials, funding for US-sponsored threat-reduction programmes in Russia will probably remain relatively static. In its first budget, the Bush administration proposed reducing spending on controlling Russian nuclear weapons and materials by 20%.[10] In 2004, the DOD sought to cut CTR spending by 10% to free up funds for other priorities, such as the war in Iraq.[11] The legacy of several fiascos – including flawed Russian and US managerial practices that wasted almost $200 million in projects at Votkinsk and Krasnoyarsk – continues to dampen enthusiasm for threat-reduction programmes in the US Congress.[12] During the 1990s, moreover, the Russian government failed to provide promised contributions to several CTR projects, requiring Washington to increase its own support to ensure their completion. Russian auditors have found that government employees misappropriated hundreds of millions of dollars of Western aid intended for nuclear dismantlement.[13] Partly as a result, most foreign threat-reduction funds now flow directly to private Russian and Western contractors, bypassing the government. Despite improvements since the turbulent Yeltsin era, Russian institutions still often have only limited capacity – in terms of competent programme managers, qualified subcontractors and suitable facilities – to use threat-reduction funds effectively.[14] Finally, the growing diversity of US programmes has engendered concerns in Congress about 'mission creep'.

In some cases, changing the way Americans provide assistance might make sense. In 2004, the chairman of Russia's state commission for chemical-weapon disarmament complained that Russia received only 30% of the funds allocated to the disposal of its chemical arsenal. The other 70% purportedly went to American entities that administered and monitored these activities.[15] An earlier study by the US General Accounting Office (GAO, since renamed the Government Accountability Office) of the Nuclear Cities Initiative, which was intended to promote civilian employment among former Soviet weapon scientists, found the same spending distribution.[16] The House Appropriations Committee has chastised DOE for giving insufficient priority to increasing the proportion of threat-reduction funds actually expended within Russia.[17] Besides the need to rebalance spending, experience suggests several other ways to improve Russian–US threat-reduction programmes.

Increase access and reciprocity

Russians complain that, in the process of helping to store, move and dismantle their excessive WMD stockpiles, Americans gain insights into Russian military practices that are not reciprocated. Formal arms-control agreements typically include verification measures that guarantee the parties roughly equivalent access for inspection and monitoring purposes. The current bilateral threat-reduction framework does not give Russian personnel the same level of access to US programmes and facilities because the Russian government does not pay for these activities. As a result, some Russians see the relationship as a pair of one-way streets: money flows into Russia from the United States, and valuable data leaks out in reverse.[18] In fact, US officials may have indirectly amplified Russian concerns by indicating that they plan to rely on the CTR programme to help verify the bilateral 2002 Strategic Offensive Reductions Treaty.[19] This unequal access leads Russians to fear that Americans learn more about Russian WMD policies and programmes than Russians know about US activities.

These concerns have encouraged Russians to restrict US access to sensitive WMD sites, especially to the 'closed cities' officially off-limits to foreigners. The Soviet Union established dozens of these so-called Closed Administrative-Territorial Entities (*zakritie administrativno-territorial'nie obrazovaniya* (ZATO)) to enhance the security and supervision of Soviet personnel developing WMD.[20] Despite a September 2001 agreement between the US DOE and the Russian Ministry of Atomic Energy (Minatom), by January 2003 DOE personnel still had not gained access to almost three-quarters of the buildings associated with Russia's nuclear-weapons complex.[21] Negotiations on a transparency protocol to permit US personnel to monitor that the Fissile Material Storage Facility (FMSF) at Mayak contains only the agreed type of fissile material – weapons-grade (*oruzheyniy*) enriched uranium or plutonium from dismantled warheads – have dragged on for over a decade.

Russian laws and regulations restricting foreign access to sensitive sites can create problems because US legislation often stipulates that American managers can spend funds only at locations that they can visit. The GAO concluded that limitations on US access to Russian sites have impeded progress in many areas.[22] Cases have arisen in which, even after Moscow authorised a visit, suspicious civilian and military authorities at the regional or local level have blocked the inspection.[23] For their part, Russians have sometimes voiced suspicions that American personnel seek more sensitive information and visits than are required merely to ensure the proper management of US-funded programmes – implying that their true intent is intelligence gathering. The sensitivity of this issue became evident in February and April 2005, when the Russian media erroneously reported that Russian negotiators had granted US inspectors access to sensitive nuclear facilities. Outraged nationalists accused the government of surrendering control of Russia's nuclear deterrent to foreigners.[24]

A persistent source of tension between the two countries has been Russian officials' refusal to grant US personnel access to the four biological-research facilities run by the Ministry of Defence. When this issue first arose in 1994, it led to the collapse of an incipient trilateral process of information sharing and reciprocal visits among American, British and Russian biological-weapons (BW) experts.[25] Since then, US officials have repeatedly expressed concern about safety and security at Russian sites, Moscow's failure to disclose all its Soviet-era facilities and stockpiles and their suspicions that some Russian institutions maintain the capacity to conduct offensive BW research, and even mass-produce biological weapons if needed.[26] The Russian government has refused to share with US scientists pathogens created by Soviet researchers, and has lobbied other FSU governments to refrain from sharing their copies of these strains.[27] In August 2005, FSB Director Nikolai Patrushev said that his agents were attempting to counter terrorist efforts to 'obtain access to biological, nuclear and chemical weapons' in the FSU, an awkward statement since international agreements required these countries to eliminate their biological weapons years ago.[28] US Congressional proposals to establish a Global Pathogen Surveillance network make Russia's participation conditional on US certification that it has no offensive BW programme.[29] Access and other difficulties have prevented the DOD from negotiating a separate CTR agreement to cover biological weapons in Russia. Although some activities are implemented through the International Science and Technology Centre (ISTC) in Moscow, a multilateral body that provides research grants for former Soviet weapon scientists, the lack of a formal bilateral accord has limited the kinds of programmes the DOD can pursue.[30]

The US State Department has expressed similar concerns that the Russian government may not have declared all of its chemical-weapon (CW) stockpiles and facilities. The Chemical Weapons Convention (CWC) – which bans the development, production, stockpiling, transfer and use of chemical weapons – requires Russia to eliminate all its agents and production facilities by 2012. Moscow must also declare all CW-related activities undertaken since the CWC entered into force in 1997. In August 2005, the State Department affirmed: 'The United States judges that Russia is in violation of its CWC obligations because its CWC declaration was incomplete with respect to declaration of production and development facilities, and declaration of chemical agent and weapons stockpiles'.[31] The assertion by several former Soviet experts that the Soviet Union had developed a new generation of chemical agents nicknamed 'Novichoks', designed explicitly to circumvent foreign detection and defences, has aroused particular concern. Russian officials have rejected US proposals that American experts be allowed to conduct short-notice visits, with unimpeded access, to suspected CW sites. Russia has permitted foreign visits only to declared storage and destruction facilities, and insists that there is no reason to conceal its holdings because it needs assistance in eliminating them. Another reason for the refusal might have been that US officials rejected Russian demands that any short-notice inspections should be reciprocated.[32]

After years of American complaints, the Russian Defence Ministry in February 2003 signed a protocol allowing US personnel unprecedented access to validate vulnerability assessments and design comprehensive security upgrades at Russia's nuclear-weapon storage sites.[33] By 2005, the two countries had worked out arrangements for US-funded upgrades at almost all Russian nuclear facilities containing weapons-usable material, except for two large complexes that assemble and disassemble nuclear weapons.[34] These serial production (*seriynoe proizvodstvo*) enterprises are perhaps Russia's most sensitive nuclear sites. Elsewhere, some progress has resulted from the greater use by Russians of lists of pre-approved visitors, and a greater reliance by Americans on Russian subcontractors and trusted agents.[35] In June 2005, Russian officials gave their US counterparts a list of 25–30 nuclear-warhead storage sites where inspections would be allowed on three occasions: at the beginning of a project, during its implementation and upon its conclusion.[36] According to National Nuclear Security Administration (NNSA) Director Linton Brooks, however, Russian officials continue to resist some American requests for access to Russia's nuclear installations.[37]

One obvious way to satisfy Russian concerns about reciprocity of access would be to grant Russian representatives more opportunities to visit WMD-related sites in the United States. When observers from 17 NATO countries and NATO headquarters were permitted to witness a Russian exercise, *Avaria 2004 (Accident 2004)*, which simulated terrorist attempts to seize nuclear weapons in transit, their presence was conditional on Russia receiving a reciprocal invitation to attend similar NATO exercises.[38] At a press conference in December 2004, Bush acknowledged that: 'I think one of the things we need to do is to give the Russians equal access to our sites, our nuclear storage sites to see what works and what doesn't work, to build confidence between our two governments'.[39] In November 2004, the DOE invited a senior delegation from Russia's Federal Atomic Energy Agency (Rosatom) to visit several sensitive US sites, including the Pantex nuclear-weapon plant, the Savannah River site and the Sandia National Laboratory, to discuss best practices for securing nuclear materials. At Pantex, Rosatom security chiefs were shown all the equivalent areas that American personnel wanted to inspect at Russian facilities.[40] The United States also invited Russian observers to witness an April 2005 nuclear-weapon security exercise at an Air Force base in Wyoming. Igor Valinkin, the commander of the Russian Defence Ministry's 12th Main Directorate, which is responsible for the physical security, transport and maintenance of Russia's nuclear arsenal, praised these exchanges for deepening both parties' understanding of how to secure their weapons.[41] Valinkin added that Russian units were adopting several US security practices.[42] Rosatom Director Aleksandr Rumyantsev has even floated the idea of creating joint Russian–US units to provide security at both countries' nuclear facilities.[43]

Besides granting Russian representatives greater access to US and NATO nuclear-weapon sites, American officials could hire Russian firms or laboratory personnel to help dismantle excessive US WMD stockpiles. Industrial partnerships could team American principal contractors with Russian subcontractors in the same way that these entities undertake CTR projects in Russia.[44] In most cases, the US government hires American companies to provide threat-reduction technology and services. These firms in turn employ Russian subcontractors for much of their work in Russia. The February 2005 Bratislava summit declaration, which stresses both governments' commitment to cooperation on 'enhancing the security of nuclear facilities in our *two* [emphasis added] countries', could justify this enhanced reciprocity.

American officials might also provide their Russian counterparts with additional financial and other data concerning US-funded threat-reduction projects. Although a GAO study in June 2005 concluded that DOD representatives were meeting more frequently with their Russian subcontractors (in part better to define American expectations and Russian responsibilities), Russians were still complaining about insufficient transparency regarding the expenditure of US funds in Russia, and US programme-evaluation criteria and systems.[45] Access to such data might encourage Russian officials to be more forthcoming with information about their own WMD-related activities. Besides details regarding Russia's chemical-weapon stockpiles or biological-weapon infrastructure, additional information about current and former Russian WMD personnel could prove especially useful for improving programmes designed to discourage them from sharing their skills or expertise with foreign governments or non-state actors.[46] Mechanisms might be created to allow scientists not currently affiliated with an institution to apply for foreign grants. The two governments could even conduct joint, or at least parallel, audits of ongoing bilateral threat-reduction programmes, as well as similar reviews of completed projects. These assessments could involve both executive- and legislative-branch personnel.

Showing more flexibility in the issuing of US visas to Russian WMD experts might also encourage reciprocity. The 'Visa Mantis' screening programme, which is designed to block entry into the United States of foreigners who might illegally try to export sensitive technologies, combined with the anti-terrorist provisions in the USA PATRIOT Act, has led to the denial or delay of visas to hundreds of Russian scientists who had been invited to visit US nuclear laboratories for various threat-reduction activities.[47] Although steps have been taken to improve the programme's administration, Visa Mantis continues to impede the realisation of US non-proliferation goals.[48]

Another solution to the access problem might be to exploit improvements in monitoring technology to enable a reduction in the American presence at Russian sites. In January 2000, the DOE decided that, if Russian officials refused to grant physical access to US personnel, they could use alternative means, such as photographs, videotapes and written certifications by site directors, to identify nuclear materials needing protection, and to design and supervise the installation of security systems.[49] Since then, American programme managers have been receiving photographs and other indirect information about the use of US-funded equipment for some projects that both sides agreed were too sensitive for direct observation by US government personnel. The Russian government has started permitting greater technical monitoring at its CW storage sites, in return for reductions in the number of foreign inspectors operating there.[50]

Expanding market incentives
Recent experience offers clear examples of the power of market-based incentives to influence Russia's non-proliferation policies. Perhaps the most prominent instance involves the sharply contrasting results of efforts to reduce Russia's stocks of highly enriched uranium (HEU), which can profitably be converted for commercial use, and its stocks of plutonium, which most American analysts consider an economic burden rather than a source of potential wealth.

Russia and the United States signed a 20-year HEU–LEU Purchase Agreement (*Kontrakt VOU-NOU*) in 1993. Under what is known as the 'Megatons to Megawatts' programme, Russia agreed to 'blend down' (by mixing with natural or depleted uranium) 500 tonnes of HEU from dismantled nuclear warheads into low-enriched (*nizkoobogashchennoe*) uranium (LEU) fuel. Russia then ships the LEU, which cannot be used for manufacturing nuclear weapons, to US civilian nuclear plants. The programme began in 1994 with the signing of a commercial agreement between the privately owned US Enrichment Corp (USEC), which serves as the US executive agent under the agreement, and the Russian government company Techsnabexport (TENEX). As of 30 September 2005, the USEC had purchased over 250 tonnes of weapons-grade HEU, the equivalent of approximately 10,000 nuclear warheads, recycling it into over 7,000 tonnes of LEU fuel. The programme has provided sufficient fuel to power about half of the electricity-producing commercial nuclear reactors in the United States.[51]

Russia receives about $500m a year from such sales, amounting to about 40% of Rosatom's annual revenue.[52] By 2013, it will have earned an estimated $7.5–$8bn under the agreement. This sum will probably substantially exceed the financial assistance provided to Russia through all other US government programmes combined. Already, it has generated more work for Russian nuclear-industry employees than all targeted US government job-creation programmes.[53] Russian officials use this money to fund research on improving nuclear safety and civilian nuclear power within Russia, and (more covertly) to subsidise the construction of new nuclear plants in China, India and other countries.[54] The substantial benefits that Russia gains from the HEU–LEU Agreement partly explain why officials have agreed to negotiate a separate transparency arrangement and other measures to reassure American taxpayers that the fuel actually comes from former nuclear warheads. The programme's success has led the Swedish Nuclear Power Inspectorate to urge European governments to consider undertaking a similar programme.[55]

The contrast with international efforts to dispose of surplus weapons-grade plutonium is stark. In September 1998, Russian President Boris Yeltsin and his US counterpart, Bill Clinton, agreed that their governments would each eliminate 34 tonnes of surplus plutonium in a parallel process by converting it to a form unsuited for weapons. Progress on implementing the resulting Plutonium Disposition Agreement (*Soglashenie ob utilizatsii izlishkov oruzheynogo plutoniya*), signed in September 2000, has however been stymied by the absence of a commercial market for Russian plutonium, and disagreements over liability provisions for accidents or sabotage.[56] The agreement requires both countries to construct plants by December 2007 to manufacture proliferation-resistant mixed-oxide (MOX) fuel for use in specially retrofitted nuclear reactors. It further stipulates that they will dispose of two tonnes of plutonium each subsequent year.[57] Another threat-reduction project, the DOE-led Elimination of Weapons-Grade Plutonium Production programme, involves primarily US-financed efforts to replace Russia's three remaining reactors producing weapons-grade plutonium (in Seversk and Zheleznogork) with conventional fossil-fuel plants.[58]

The two sides' diverging appraisals regarding plutonium's future value in civilian energy production have substantially complicated this issue. Representatives of the Russian government and nuclear industry have traditionally considered plutonium a potentially profitable fuel source, in the form of MOX and possibly thorium-based fuel, and for a future generation of fast-neutron breeder reactors (*reaktori na bistrix neytronax*) that would both burn and produce plutonium. They expect these technically and financially challenging power systems to become increasingly viable as the technology improves and the global supply of natural uranium declines. Russia therefore retains a closed nuclear fuel cycle that uses chemical separation to reprocess spent nuclear fuel and separate the plutonium and reusable uranium from radioactive waste. Russian officials also rejected proposals to 'immobilise' excess weapons-grade plutonium, for instance by mixing it with radioactive waste or a special type of sand, rather than recycling it as fuel for civilian reactors.[59] In contrast, US experts have traditionally believed that plutonium-based fuel cycles will not become technically or commercially viable for decades, if ever. Besides minor experiments, the United States has since the 1970s refused to separate plutonium from reprocessed spent nuclear fuel because of the proliferation risks. Instead, the United States relies on a once-through fuel cycle, in which spent nuclear fuel is placed in storage.[60]

Although Russian analysts might be proved more prescient in the long run, the short-term commercial prospects for Russia's plutonium are bleak. It will cost some $2bn to dispose of the stipulated 34 tonnes of excess Russian weapons-usable plutonium.[61] MOX fuel is much more expensive than uranium fuel alone. European countries have had to subsidise its production to promote its commercial use. The world's supply of natural uranium has remained sufficient to meet the limited increase in demand for nuclear power, while the difficulties of spent-fuel reprocessing have persisted. Meanwhile, the net global supply of separated (unirradiated) plutonium potentially available for use as MOX fuel in civilian light-water reactors has been growing by some ten tonnes annually since 1997.[62] Russia and the United States still have an estimated 150 and 100 tonnes, respectively, of surplus weapons-grade plutonium.

Efforts by US programme managers to employ Russian special-purpose equipment and local firms as subcontractors provide another example of a successful market-based approach to threat reduction. For example, CTR managers have hired Russian shipyards to dismantle decommissioned Russian SSBNs. With US government approval, several Russian firms have used CTR-provided equipment to dismantle other types of decommissioned Russian nuclear-powered submarines under Russian or non-US programmes. This flexibility helps to reduce costs, develop local constituencies for projects and transfer marketing, management and technical skills.[63] A GAO report has said that DOE 'project teams are designing systems that use equipment produced in Russia rather than foreign-made equipment because Russian equipment may be easier for the sites to service and replacement parts may be more readily available'.[64] Problems have arisen, however, when non-market considerations have interfered with this process. For example, Russian efforts in 2003 to select a preferred contractor at the planned Shchuch'ye Chemical Weapons Disposition Facility (CWDF) delayed construction for months and increased the project's costs.[65] More positively, Russian firms that have developed profitable commercial ties with US companies have become more concerned about not violating Russian or foreign export controls (for example, by selling prohibited technologies to Iran).[66]

Clearly, market-based solutions are not applicable in all cases. The widespread participation in the multinational A.Q. Khan network, whose shady dealers for years illegally sold nuclear-weapon technology and expertise to rogue regimes, demonstrates the power of the profit motive (as well as ideology and conceit) to motivate black-market nuclear entrepreneurship. For some people, the limited funds provided by any conceivable programme designed to redirect scientists towards peaceful pursuits, or at least prevent them from selling their expertise to foreign agents, cannot counter the attraction of such potential lucre. Instead, non-market barriers (administrative, geographic or physical) and disincentives (cultural, professional or social) must operate to outweigh financial incentives. These non-market factors clearly have played a role in Russia's case. Few Russian scientists, engineers or other workers in WMD-related fields appear to have attempted to sell their access or expertise, despite their possible black-market value.[67]

Market considerations have also worked against accelerating HEU purchases from Russia. Under current arrangements, the amount of HEU blended down is determined, not by security concerns or the capabilities of facilities, but by US anti-dumping legislation designed to protect domestic uranium suppliers.[68] In June 2002, Russia and the United States had to authorise USEC and TENEX to adopt a new pricing formula that took into account short- and long-term changes in uranium's market price. The revised contract, which aimed to limit disruptions from short-term price swings and end the need for USEC to pay more for Russian nuclear fuel when less expensive alternatives existed, will reduce the revenue Russia can expect to receive through the remainder of the agreement (until 2013) by hundreds of millions of dollars.[69] One quasi-market measure to encourage increased HEU purchases from Russia would be for the American government to buy more than the stipulated 30 tonnes of HEU per year, have Russia blend it down below weapons grade, but place the excess in storage until market conditions warrant its use as nuclear fuel.[70] If Russia carries out its plans to expand its domestic nuclear-power capacity, demand for additional fuel supplies could rise markedly in the coming years.

Another problem area has been attempts to convert former Soviet biological-warfare scientists and their laboratories into self-sustaining commercial entities selling non-military products, such as vaccines, medicines and other drugs. Several US programmes have financed these conversion projects. Multilateral support has come from both the G-8 Global Partnership and the ISTC, which funds commercial projects, as well as business training and related activities to promote the transition of former weapons scientists into non-military employment.[71] Private contributions have also become increasingly important. During the 2004 fiscal year, the DOE's Global Initiatives for Proliferation Prevention (GIPP) Programme obtained 60% ($24m) of its total project funding from non-governmental sources. Private contributions were especially valuable in cases where American law prohibited the use of US government funds.[72]

These conversion efforts have encountered several deep-rooted problems. Most ISTC research grants have had a short-term focus, and topics have been selected on the basis of what the technology permits, rather than what the market wants. They have indirectly subsidised weapons research and have attracted little interest in their findings. Above all, only a handful of former Soviet BW scientists have commercialised their research projects or embarked on entirely new, non-defence careers. Other recurring problems include the limited entrepreneurial experience of ex-Soviet scientists, sluggish responses to potential sales opportunities, a lack of established markets or contacts in foreign countries and an inability to compete with the extremely high standards (for decontamination, purity and quality control) found in most Western biotechnology laboratories.[73] Finally, while foreign firms would like to exploit mutually profitable business opportunities, they do not want to create competitive rivals.[74]

Russian and Western governments should acknowledge these difficulties and redirect their efforts to inducing these scientists to work on biosafety and biosecurity (*biobezopasnosti*), rather than focusing on disarmament.[75] Russian help in this area would be especially valuable since US and other international threat-reduction programmes have devoted fewer resources to securing and countering dangerous biological pathogens than they have to nuclear and chemical threats. Appropriate topics could include safeguarding food supplies, force protection, enhanced public-health surveillance (which alerts governments to both natural and man-made biological threats), and developing counteragents. Questions relating to liability and environmental protection, anticipated profit margins and licensing and permit policies severely impede BW defence research in Western countries.[76] Russian regulations are more flexible, health authorities tend to approve projects more rapidly and the public seems more tolerant of having dangerous biological agents in the neighbourhood.[77] Working on biodefence projects would also more closely dovetail with these scientists' past endeavours and core competencies. Many Soviet specialists worked on protecting civilians from biological weapons (observers estimate that Russia still enjoys a 20-year lead over America in this area).[78] Providing support for research on biodefence might also attract experts working at Russian Defence Ministry facilities. Thus far, they have stood aloof from international threat-reduction efforts, but they might want to participate in biodefence projects that could strengthen Russian as well as Western force-protection capabilities.[79] The BW case suggests that, while some elements of Russia's weapons complex can and should be reduced through economic diversification, other components can best contribute to shared Russian and US goals by continuing to work on defence-related projects, even if foreign funding is required.[80]

Enhancing Russian support
Some American critics of threat-reduction spending believe that it serves primarily as a foreign-aid programme, providing US money for what should be Russian responsibilities. At worst, they fear that it allows Moscow to redirect resources towards strengthening its own armed forces, though it is doubtful whether, in the absence of CTR, Russian offi-

cials would choose to fund threat-reduction activities rather than military modernisation. In the past, the Russian government has failed to fund its agreed share of the costs of some threat-reduction projects.[81] While critics have called on Russia to devote more of its growing economic resources to such programmes, even strong supporters of bilateral threat reduction advocate transforming the relationship 'from patronage to partnership'.[82] These groups argue that Russian officials should provide additional funds and redouble efforts to reduce obstacles to threat-reduction programmes within Russia. In return, Russian policymakers would be more closely involved in programme planning and management; there could be more joint coordinating committees, for instance.[83]

An increase in Russia's contributions to threat-reduction activities would have several beneficial effects. First, it would reduce foreign scepticism regarding Russian commitment to the enterprise, thereby weakening opposition to providing non-proliferation assistance. Secondly, it would help to elevate Moscow's status from that of a resentful supplicant to a genuine partner in a common endeavour. The current relationship engenders resentment in both countries. Russians often feel manipulated and, when their non-financial, 'in kind' contributions are not adequately recognised, unappreciated. Americans sometimes simply feel exploited. Thirdly, programme implementation might improve if Russia's role in designing and operating threat-reduction activities increased, to correspond with its elevated status and contribution.

The Russian government has made some moves to increase its contributions. The revival of the economy that began in the late 1990s has enabled the authorities to pay the salaries of security guards, scientists and other employees at WMD-related facilities on time, and to fund other safety and security measures.[84] Russia plans to spend at least $2bn on threat-reduction activities during the ten-year period encompassed by the 'Global Partnership Against the Spread of Weapons and Materials of Mass Destruction'. Launched at the June 2002 G-8 summit in Kananaskis, Canada, the Global Partnership provides for the enhanced coordination of national programmes relating to WMD non-proliferation, counter-terrorism and nuclear safety. Moscow has also changed the way it funds CW elimination, so that financing can be increased if external support falls short of expectations.[85] In 2005, the federal budget doubled allocations for CW destruction (to almost $400m).[86] Russian officials recognise that such spending creates jobs, and that a failure to fulfil the CWC's requirements could subject Russian chemical companies, which export billions of dollars' worth of fertilisers annually, to international trade sanctions. All parties acknowledge, however, that Moscow alone cannot afford to fund the complete elimination of its stockpile by 2012. Russia also requires foreign aid to cover the estimated $4bn cost of dismantling its decommissioned nuclear submarines, moored precariously along its coasts.[87]

Russian officials should consider increasing support for other high-priority areas. Defending against biological terrorism represents a good example. In early 2005, Lev Sandakhchiyev, director of the influential Vektor State Science Centre of Virology and Biotechnology in Novosibirsk, complained that the Russian government had ceased funding research into BW countermeasures. According to Sandakhchiyev, Russia, Europe and the US therefore had no 'real, constructive programs' for cooperating against bioterror threats.[88] None of the other G-8 governments is spending heavily on collaborative projects in this area either.[89] Another issue warranting more Russian spending is plutonium disposition. Citing Russia's substantial 'in kind' contributions of territory, labour and technologies, Rosatom Director Rumyantsev has insisted that foreign governments should cover the entire estimated $2bn cost of implementing the 2000 Plutonium Disposition Agreement. The United States and other potential foreign donors have made clear their belief that the Russian government should make at least some financial contribution.[90] A compromise might allow Russia to provide funds using some of the proceeds it expects to earn from selling MOX fuel or space at a possible International Spent Fuel Storage Facility located in Russia. Russian financing could help to increase the number of automated radiation detectors constructed in Russia under the auspices of the DOE's worldwide Second Line of Defense Program. Limited US funds have allowed DOE to install sensors at only about 60 of Russia's most sensitive seaports, airports and border crossings.[91]

Besides spending more, the Russian government should show greater responsiveness towards foreign threat-reduction activities on its territory. Russian visa, customs and tax policies continue to disrupt projects. Russian officials need to accelerate their plans to exempt US and other foreign threat-reduction activities from all taxes and customs. Other governments insist that funds they allocate to threat reduction should not indirectly subsidise Russian programmes that they have not explicitly consented to support. Although the central government has agreed to excuse foreign programmes from taxes, securing such exemptions at the local level has in many cases proved exceedingly difficult. Project managers can often recover payments, but only after much time and effort. Russian (and US) officials also need to make it easier for foreigners employed on major projects to obtain long-term, multiple-entry visas.[92] American contractors working in Russia must frequently leave Russian territory to apply for renewal of their 90-day visas.[93]

Russia and the United States also urgently need to renew the so-called CTR Umbrella Agreement of 1992, which was provisionally extended in June 1999 for seven years.[94] Such framework agreements, negotiated by the State Department, provide the *de facto* legal basis for almost all American-funded threat-reduction projects. The accord with Russia grants US personnel a comprehensive set of protections, exemptions and rights, including freedom from taxes and customs, various privileges and immunities and the right to verify that any assistance is used only for its intended purpose. According to the then Deputy NNSA Administrator Paul Longsworth, failure to renew the Umbrella Agreement would require halting all US threat-reduction activities in Russia.[95]

The protracted dispute over the provisions protecting US contractors from liability in case of accidents or other disaster (*vopros yuridicheskoy otvetstevnnosti storon*) during their work in Russia has been the most problematic issue affecting the Umbrella Agreement. The high stakes at issue – a nuclear catastrophe could cause hundreds of billions of dollars of damage – have discouraged compromise. The US government has pushed for maintaining the generous provisions of the June 1999 extension protocol, which both sides have applied despite the failure of the Russian parliament (the Duma) to ratify it. Its liability provisions provide for almost complete indemnification for Americans involved in threat-reduction projects in Russia, leaving the Russian government in principle solely and unconditionally liable. Russian officials, however, have insisted that future bilateral arrangements must follow the provisions of other international arrangements in this area, such as the Vienna Convention on Civil Liability for Nuclear Damage (which Putin has signed into Russian law) and the May 2003 Framework Agreement on the Multilateral Nuclear Environmental Programme in the Russian Federation (MNEPR). The MNEPR governs bilateral projects in northwest Russia, including decommissioning nuclear submarines, managing radioactive waste and enhancing the safety and security of nuclear reactors. The MNEPR's separate Protocol on Claims, Legal Proceedings and Indemnification, which the US government has refused to sign, allows for arbitration in liability disputes between Russia and the other signatories, and holds actors legally responsible for any intentional wrongdoing. Russian representatives want to apply these provisions to US contractors, making them or the United States legally liable for damages caused by American-hired entities in Russia from any 'premeditated act' ('*prenamerennoe deystvo*').[96] US officials reject these provisions, claiming that they provide insufficient safeguards against the vagaries of the Russian legal system.[97]

Disputes over liability protection for US contractors led in 2003 to the non-renewal of the 1998 Plutonium Science and Technology Agreement (a five-year accord entailing joint concept design, research and development and small-scale pilot projects to test plutonium-disposition methods), and the suspension of new projects under the 1998 Nuclear Cities Initiative.[98] Since then, both governments have feared that any liability standards they accept for a specific programme or area could establish an unfavourable precedent for negotiating the Umbrella Agreement.[99] Members of the US Congress, non-proliferation analysts and executives in the nuclear industry have all urged the two governments to adopt a more flexible approach. Some have offered creative proposals for overcoming the dispute.[100] In July 2005, Russian and American negotiators reached a deal in principle regarding liability provisions for the Plutonium Disposition Agreement, but American negotiators have said that any resolution of this issue would not necessarily establish a precedent for settling other liability disputes, including the Umbrella Agreement.[101]

Integrating programmes
For many years, analysts, policymakers and members of the US Congress have advocated designating a single senior non-proliferation official in each government to coordinate their respective programmes. This person, enjoying direct access to the president and with influence over expenditure in this area, would work exclusively to improve Russian–US threat-reduction programmes.[102] Such restructuring could help to overcome impediments to cooperation that transcend government departments, such as the disputes relating to access, liability and visas. At present, only direct intervention at the presidential level can solve many of the complex interagency issues for which no single actor has the authority or resources required to force a decision.

At their Bratislava summit in February 2005, Bush and Putin announced the establishment of a bilateral Senior Interagency Group (SIG) for Cooperation on Nuclear Security, chaired by the US Secretary of Energy and the Director of Rosatom, to oversee implementation of the summit initiatives on nuclear-security cooperation. The SIG has developed a Joint Action Plan for security upgrades at Rosatom and Russian Ministry of Defence facilities, as well as 'prioritised timelines' for the repatriation of HEU fuel from foreign countries to Russia and the United States. It will deliver reports on these issues to the Russian and American presidents every six months. The SIG is also organising bilateral workshops on sharing best practices and promoting a 'security culture', and is conducting an exercise on managing the consequences of nuclear incidents.[103] Although the SIG will help to identify problems, actually resolving disputes or exploiting opportunities will require continued presidential intervention and an institution with greater bureaucratic influence.[104]

In a January 2005 report, the GAO found that US threat-reduction efforts in Russia (and other countries) suffered from coordination problems because several government agencies were pursuing similar programmes. For example, GAO analysts could not identify any authoritative government-wide guidance delineating the roles and responsibilities of the US agencies managing border-security programmes in the FSU. The report also stressed the need for the National Security Council, which currently coordinates US threat-reduction and non-proliferation programmes, or another body to develop and enforce an integrated strategy across the executive branch.[105] The growing involvement of the Department of Homeland Security in WMD-related issues increases the need for stronger coordination. The acting director of the new Department of Homeland Security Domestic Nuclear Detection Office, Vayl Oxford, has asserted its responsibility 'for developing an overall global architecture that assesses and links these [DOD, DOE and State Department] programs in an effort to ensure that the nation proceeds with a single, comprehensive prevention and detection strategy'.[106] The GAO is now investigating possible redundancies and inadequate information-sharing among the US government bodies responsible for detecting illicit trafficking in nuclear materials. Media reports indicate that the federal agencies involved in BW defences also suffer from overlapping jurisdictions.[107]

On the Russian side, having a single person in charge of all threat-reduction activities could provide greater stability in an area that has repeatedly been affected by major government reorganisations. In 2003, a group of Russian non-governmental experts concluded that 'the weakest point of the Russian mechanism of cooperation with other countries in eliminating the Cold War legacy is the lack of a "super-agency", a coordinating and controlling body capable of overcoming departmental self-interest'. The experts added that a vice-prime minister would need to head such a super-agency for it to be effective.[108] In 2003, Putin made a deputy prime minister responsible for coordinating all threat-reduction activities in Russia, but the incumbent's ability to affect programmes and budgets remained uncertain.[109]

Repeated Russian government reorganisations, combined with frequent turnover in US government personnel, have made it difficult to develop lasting partnerships and have complicated programme coordination. For example, DOE has had to respond to several changes in the status of Russia's nuclear agencies. In March 2004, Putin decided to dismantle Minatom, incorporating most of it into a new Ministry of Industry and Energy (MIE).[110] As the government entity responsible for Russia's civilian and defence nuclear complexes, including developing and testing nuclear weapons, Minatom had been the DOE's major Russian interlocutor. In May 2004, Putin partially reversed the decision and removed Rosatom and the Federal Service for Atomic Inspection (the former GosAtomNadzor) from the MIE, placing them under the supervision of the prime minister's office.[111] The two organisations also regained the power to negotiate and sign international agreements, and to authorise customs and tax exemptions.[112]

On 28 July 2004, the Russian government issued Decree (*ukaz*) no. 316, 'The Guidelines on the Federal Agency of Atomic Energy', which makes clear that Rosatom is strictly an executive agency. Unlike the former Minatom, its task is to implement but not determine policies.[113] Around this time, Putin authorised the Russian Ministry of Defence to supervise Russia's 'nuclear weapons complex' (*yaderniy oboronniy kompleks*). The ministry's responsibilities encompass the development, production, modernisation, decommissioning and dismantling of nuclear weapons; nuclear safety; and the supervision of Russia's nuclear test site on Novaya Zemlya. Although Putin said that the complex 'fell, and would fall under his personal attention', the Defence Ministry normally defines Rosatom's specific tasks in defence-related contracts and programmes. As a result, Rosatom became the only Russian agency subordinate to two ministries.[114] American DOD officials say that the reorganisation delayed several threat-reduction programmes because of the need to renegotiate implementing agreements.[115] At present, US government agencies must work with a plethora of Russian ones: Rosatom, Rosenergoatom (which manages Russia's civilian nuclear plants), the Navy, the Federal Service for Atomic Inspection and the Ministries of Defence, Interior, Education and Economy.

Similar confusion and complexity prevail regarding the bureaucratic structures for Russia's biological- and chemical-weapon sectors. Until the recent reorganisation, nine separate government entities had jurisdiction over BW issues.[116] As for CW elimination, the abolition of the Russian Munitions Agency in 2004 has disrupted and made less transparent Russian decision-making in this area.[117] It was only in September 2004 that the government established a commission on biological and chemical security to coordinate the various Russian entities active in the BW and CW fields.[118]

To supplement (or, if necessary, substitute for) parallel and interlinked non-proliferation coordinators in both countries, the United States and Russia could revive something like the Gore–Chernomyrdin Commission. Begun in 1993, and continuing throughout the Clinton administration, the Commission's twice-yearly sessions involved many senior cabinet-level officials from both governments under the leadership of US Vice-President Al Gore and Russian Prime Minister Viktor Chernomyrdin. Subordinate working groups developed plans for future joint activities and evaluated ongoing and past cooperative projects. Each session typically produced a dozen agreements and several important joint statements. The commission proved especially valuable in helping to overcome coordination problems within the chaotic Russian bureaucracy. It also focused high-level attention in both capitals on bilateral threat-reduction issues, and helped to manage major disagreements, such as over NATO enlargement and the Kosovo war.[119]

The Bush administration dismantled the commission soon after taking office. Senior officials now discuss security issues primarily when visiting Moscow and Washington, and US officials have encouraged private-sector organisations in both countries to address non-security issues directly.[120] Reviving a high-level body like the Gore–Chernomyrdin Commission, with perhaps a narrower range of participants, would facilitate the timely and integrated analysis of the most important bilateral security issues. The body's security-related subgroups, which could incorporate existing joint working groups like the Consultative Group for Strategic Security, might provide mechanisms for exchanging information and threat assessments, and for undertaking combined planning. Another important commission function would be promoting mutual transparency regarding military technologies and weapons systems. In the future, it might even oversee joint defence research and development programmes.[121] It could also supervise the creation of a comprehensive and mutually transparent database linking non-proliferation threats and programmes. Such a matrix, which would incorporate data from both countries' intelligence communities and other sources, would guide resource-allocation decisions and produce an integrated bi-national threat-reduction strategy. Its decisions would enjoy the imprimatur of both governments' high-level representatives, facilitating their implementation even in the face of bureaucratic obstacles.

Transforming legislative oversight

Congressional interest has helped to improve US threat-reduction programmes. Not only did members of Congress launch the original CTR initiative, but several have developed an uncommon expertise in various projects. Their manifest concern has also enhanced US representatives' leverage, making their threats and promises more credible. In authorising and appropriating funds for threat-reduction activities, Congress has typically placed restrictions on expenditures (such as prohibiting spending on environmental restoration). Furthermore, since 1993 members have imposed certification requirements on potential recipients. Currently, the president must certify to Congress that the proposed recipient is committed to: (1) making a substantial investment in WMD destruction; (2) foregoing military modernisation programmes that exceed its legitimate defence requirements or involve replacing destroyed WMD; (3) refraining from using fissile materials or other components from destroyed nuclear munitions in new nuclear weapons; (4) facilitating US verification of American-funded activities; (5) complying with all relevant arms-control agreements; and (6) observing internationally recognised human rights, including the protection of minorities. In early 2002, the Bush administration determined that it could not certify Russia's fulfilment of all six general programme requirements. This decision resulted for the first time in the suspension of new funding for CTR programmes between April and August 2002, at which point Congress authorised the president to waive the certification requirement temporarily.[122]

Starting with the 2002 National Defense Authorization Act, Congress has permitted CTR funding of the Shchuch'ye CWDF, but only after the president certified that Russia had met six additional conditions: (1) provided full and accurate information regarding the size of its CW stockpile; (2) made a substantial financial contribution (at least $25m annually) to CW destruction; (3) developed a practical plan to destroy all its nerve agents; (4) changed Russian law to require the elimination of all nerve agents at a single site (i.e., Russia would rely on Shchuch'ye alone rather than construct additional billion-dollar facilities); (5) agreed to destroy or convert its nerve-agent production facilities; and (6) secured an international commitment to fund the CWDF. In the 2003 Defense Appropriations Act and subsequent legislation, Congress granted the president authority to waive these conditions for reasons of US national security. Unable to certify that all six conditions had been met in 2003 and 2004, Bush exercised this waiver, but only after lengthy work delays at the CWDF.

Besides the postponements, the certification requirements typically require government personnel to devote hundreds of hours to preparing documentation for the certification and waiver processes.[123] In addition, they tend to make the protracted negotiations with Russia less a joint endeavour between trusting partners seeking to achieve mutually beneficial goals, and more a struggle aimed at displacing risks and enforcing behaviour. The certification requirements also appear to have had little independent effect on Russian policies. For these reasons, some members of Congress have sought to eliminate them.

As an alternative to explicit certification requirements, Congress might rely more on performance-based oversight, giving the executive branch more flexibility in meeting congressionally defined goals, timetables and measures of performance. In 2003, the DOD revised its audit procedures and performance measures across all CTR programme areas to incorporate lessons learned from past projects.[124] A June 2005 study by the GAO found that the reforms had significantly strengthened DOD's management in the areas of organisational structure, risk assessments, performance measures, programme reviews, management training and communications. CTR programmes now almost always meet congressional reporting requirements.[125] To minimise unpleasant surprises, however, Congress might require executive agencies to provide more frequent risk assessments for major threat-reduction projects. Both GAO and DOD recognise that skilful management can mitigate, but not entirely eliminate, risks given all the technical and political uncertainties involved in these complex, costly and lengthy projects in countries that often have strained relations with Washington. In addition, members might want to receive briefings on the lessons-learned studies that the DOD plans to undertake of completed CTR projects.

As in other complex areas, such as homeland security, consolidating the relevant congressional oversight committees might improve the supervision of threat-reduction programmes in Russia and elsewhere. Reflecting the multiple executive-branch agencies involved in these programmes, the Senate and House Committees on Appropriations, Armed Services, Budget, Commerce, Energy, Foreign Relations and International Relations, Governmental Affairs, Intelligence and several others exercise some jurisdiction. Although the current structure generates multiple perspectives and policy entrepreneurship, it also promotes incoherence and wasted resources, especially in the time consumed by congressional hearings.

If ties between the US Congress and the Russian Duma deepen, bi-national legislative oversight of some threat-reduction programmes might become feasible. It would be useful to involve members in joint briefings, simulations and related activities to help overcome the divergent perceptions and misunderstandings that often dominate the debates in both legislatures on threat-reduction issues. They might establish a formal joint working group to institutionalise these reviews. Members of the Duma and the Federation Council (the upper house of the Russian parliament)

have long expressed a desire to participate in systematic exchanges with
their American counterparts.[126]

Conclusion

The record of the past 15 years of Russian–American collaborative threat
reduction suggests several conclusions. Inviting more Russian representatives to visit US sites, hiring Russian firms or personnel to help dismantle
excessive US WMD stocks and providing more data concerning US-
funded projects in return for more detailed information about Russian
WMD-related activities and assets could help overcome disputes over
access. Market-based incentives encourage profitable bilateral cooperation. Expanding Russian contributions to threat-reduction activities within
and outside Russia could create a more equal and durable relationship.
Restructuring the Russian and US non-proliferation bureaucracies could
enhance senior-level attention, and could improve the integration of these
programmes. Finally, joint or parallel legislative oversight might promote
better mutual understanding of threat-reduction activities.

CHAPTER TWO

Furthering Russian–US defence cooperation

The Russian and American defence communities have substantially improved their bilateral cooperation since the Soviet Union's demise. In April 2005, Secretary of State Condoleezza Rice told a Russian radio audience: 'I believe that our military-to-military cooperation is perhaps the best that it has ever been'.[1] The Russian and US militaries have conducted major joint operations, most prominently in the former Yugoslavia, and the two governments have discussed possible combined anti-terrorist operations in third countries. Their non-proliferation experts increasingly collaborate on threat-reduction projects outside Russia.

Although the Russian and US armed forces have made some progress in overcoming impediments to improved interoperability, further efforts are needed to enhance mutual understanding and to improve their ability to operate together. The Russian and American defence research, development and acquisition communities might produce innovative military technologies and weapon systems beneficial to both parties. Opportunities for effective collaboration appear especially promising in enhancing joint defences against improvised explosive devices (IEDs, or *netraditsionnie vzrivnie ustroystva*) radiological dispersal devices and other forms of nuclear terrorism and, as noted earlier, harmful biological agents. Substantial difficulties remain, however, regarding bilateral ballistic-missile defence cooperation and the alleged export of WMD- and missile-related technologies by Russian entities to states of proliferation concern.

Conducting better joint military operations

The US defence community has shown an intense interest in cultivating military-to-military contacts with its Soviet/Russian counterpart for at least two decades.[2] Promoting deeper interoperability (*operativnaya sovmestimost'*) could become even more important now that both countries are contemplating much more demanding joint military operations. These could include humanitarian relief missions in non-permissive environments, or even combined operations to secure or destroy vulnerable WMD assets that might fall under terrorist control. In 2002, Russian and US officials discussed a possible joint military operation against al-Qaeda operatives in Georgia's Pankisi Gorge.[3]

Effective multinational military operations require high levels of interoperability. Technical interoperability is enhanced when engineers, weapon designers and other defence experts develop common military technologies. More subjective forms of interoperability, such as appreciating the other party's preferred tactics, techniques and procedures, require close, frequent and sustained contact. The DOD's International Military Education and Training Program (IMET) provides training in English for Russian military officers and civilian officials in peacekeeping operations, the development of non-commissioned officers, civil–military relations and other topics.[4] However, it can take years of military engagement to reshape the deep-rooted perceptions and practices of a former adversary.

Enhanced mutual understanding could prove especially useful in Central Asia, where the two militaries operate independently, but in close proximity. In Kyrgyzstan, Russian and US forces have little formal communication despite their close military bases, a situation inadvertently highlighted in April 2005 when Defence Minster Ivanov said that 'Russian and US military bases in Kyrgyzstan are not bothering each other'.[5] Russian and US forces have deployed in other Central Asian countries as well.[6] These contingents should consider institutionalising regular consultations among base commanders, and conducting joint exercises on force protection, humanitarian relief and counter-terrorism to explore how they might interact in a crisis, which at a minimum would help avoid friendly fire and other incidents. Staff members from CENTCOM, the Joint Staff and the Office of the Secretary of Defense (OSD) could regularly brief their Russian counterparts on US military activities in Central Asia, just as EUCOM staff do regarding some US defence initiatives in the Caucasus.[7]

The Russian military remains more isolated from external influence than most other Russian institutions.[8] Involving Russian military representatives in international activities has typically required strenuous and determined efforts by senior policymakers in both countries. Moreover, Russian and US military educational exchanges have typically been one sided. Russian officers have enrolled in professional military education institutions in the United States, but no American officers have studied at Russia's military academies.[9] The continuing lack of transparency in Russia regarding military activities and plans has also complicated efforts at engagement.[10] The reactionary views of many Russian officers will probably also impede international cooperation. A leaked Defence Ministry poll found that, as of early 2005, some 80% of the officers surveyed considered Russian government policies 'too weak', and only 17% of respondents said that they trusted Putin.[11] Russian forces still conduct large-scale exercises with scenarios implying a nuclear war with the United States.[12] Moscow's brutality and ineffectiveness in Chechnya have made some US defence officials uncomfortable about working directly with their Russian counterparts.

Despite these difficulties, US attempts to engage the Russian armed forces should continue. The military will remain an important actor given its size, aggregate resources and close relations with other Russian institutions. Its budget should approximate $22bn in 2005, an enormous sum relative to Russia's GDP, and Russian defence reform is closely linked with many other aspects of the country's transformation. For its part, the American military has a unique role to play in engaging Russia's armed forces. For various reasons related to their shared Cold War legacy and the demonstrated effectiveness of US military operations, Russian defence leaders seem most comfortable working with their American counterparts, rather than with other militaries. Expanding reciprocal contacts would help to overcome the lack of understanding regarding the US military and its professional ethos which apparently still pervades the Russian armed forces.[13] Russians need to appreciate the high value that NATO militaries place on upholding human rights, curbing abuses and unprofessional conduct and treating civilian control (including effective parliamentary oversight) as more than just a measure to prevent coups. Curtailing bilateral military contacts to protest against Moscow's policies will not help to make the Russian armed forces a less hostile institution. The mix of cooperation and conflict that characterises Russian–American relations means that military-to-military contacts and other forms of bilateral security engagement are both necessary and possible.[14]

Expanding technological collaboration to promote global security
Enhancing ties between the Russian and American defence research, development and acquisition communities could produce mutually bene-

ficial military technologies and weapons systems. The two countries have outstanding and often unique expertise in many defence fields, including ballistic missiles, combat aircraft, nuclear submarines and – especially – WMD. Although financial problems limited Russian military research and development during much of the 1990s, funding and programmes have increased markedly since then.[15] Russian and US military designers often adhere to different methodologies and schools, allowing them to apply complementary perspectives to problems. The successful cooperation between Russian and US non-proliferation experts has demonstrated the value of pursuing joint proposals and technologies for common security problems. For example, the bi-national Warhead Safety and Security Exchange Agreement (WSSX) has helped to improve the safety and security of nuclear warheads by developing better radiation detectors and explosive-resistant storage units.[16] Russian and US analysts have been evaluating threats from cyber-terrorism, especially computer networks attempting to seize control of nuclear weapons or mislead early-warning and command, control and communications systems into launching unauthorised missile strikes.[17] The time appears ripe to fund demonstration projects of the most promising of these concepts.

American officials clearly appreciate the value of obtaining access to Russian defence technologies. For example, two core objectives of the Cooperative Biological Research project of the US Biological Weapons Proliferation Prevention programme, which engages with Russia's former BW scientists, are to provide 'US access to this scientific expertise to enhance preparedness against biological threats' and 'opportunities for transfer of especially dangerous pathogens for additional study in the US to improve public health and for forensics reference'.[18] In addition, one of the agreements announced at the Bratislava summit involves sharing information between Russian and US experts on the kinds of improvised explosive devices that have so bedevilled American counter-insurgency operations in Iraq.

At their May 2002 summit in Moscow, Bush and Putin established a US–Russia Working Group on Advanced Nuclear Technologies to explore nuclear-reactor and fuel-cycle technologies that could produce less radioactive waste and better consume stocks of existing weapons-usable fissile materials.[19] Although little came of the initiative, Russian specialists remain interested in multilateral programmes to develop commercial nuclear-energy technologies and fuel cycles that could prove more proliferation-resistant than existing procedures.[20] Russian experts initiated the International Project on Innovative Reactors and Fuel Cycles, and have collaborated in the related US-led Generation IV International Project. Russian scientists are discussing with their US counterparts whether floating nuclear power plants (*plavuchie atomnie elektrostantsii*) might entail fewer proliferation problems than fixed land-based facilities because, under 'turnkey' arrangements, recipient countries would have to return the plant to its country of origin after it had ceased operations.[21]

Until now, disputes about Russia's nuclear cooperation with Iran have limited formal cooperation on these projects. Nevertheless, opportunities for Russian–US collaboration on joint or multilateral threat-reduction projects outside of the FSU increased substantially in June 2003, when the G-8 governments decided to expand the scope of the Global Partnership. The US administration has pledged $10bn to the initiative over a ten-year period, and the other G-8 members have promised a comparable amount. Since 2003, more than a dozen other countries have joined the programme, collectively pledging over $250m.[22] Many contributions focus on areas of special interest to the country concerned. For example, Norway has financed the dismantling of decommissioned Russian nuclear submarines in the Barents Sea. In the past, G-8 Partners have helped to support infrastructure projects that US law prohibited American officials from funding.[23]

The Russian government has pledged to contribute $2bn to the Global Partnership, primarily to help eliminate Russia's chemical weapons and dismantle its nuclear submarines. Russian officials have also agreed that foreigners involved in Partnership projects will enjoy the same privileges regarding access, taxes and other issues that Americans receive through their bilateral threat-reduction programmes. Despite initial concerns that including additional countries would reduce the funds available for use within Russia, they have also endorsed expanding threat-reduction activities in other countries, such as Ukraine, provided that projects addressing Russian concerns remain a priority.[24] At their July 2005 summit in Gleneagles, UK, the Global partners made clear 'the Partnership's openness in principle to further expansion in accordance with the Kananaskis documents, and in the context of the ongoing focus on projects in Russia'.[25]

Another opportunity for joint Russian–US threat-reduction projects beyond Russia arose in May 2004, when US Secretary of Energy Spencer Abraham announced a Global Threat Reduction Initiative (GTRI) to identify, secure and dispose of stockpiles of vulnerable nuclear and radiological materials and related equipment at civilian sites throughout the world. The GTRI has four core elements.[26] The first two consist of accelerated DOE efforts to repatriate Soviet/Russian and US-origin HEU from foreign countries. The returned HEU includes both 'fresh' (i.e., unused) and spent nuclear fuel. Current plans are to complete the repatriation of all Soviet/Russian-origin fresh nuclear-reactor fuel by the end of 2005, all Soviet/Russian-origin spent fuel by 2010, and all US-origin spent fuel for research reactors by 2015. The third element, the Reduced Enrichment for Research and Test Reactors (RERTR) programme, funds efforts to convert the cores of selected civilian research reactors worldwide, many of which are Soviet built, to use LEU rather than HEU fuel. (Experts estimate that approximately 20 tonnes of HEU exists throughout the globe as fuel for over 130 civilian research reactors.[27])

The fourth GTRI element, the International Radiological Threat Reduction programme, involves identifying and securing nuclear and radiological materials and related equipment not addressed by current threat-reduction activities. According to the GAO and the International Atomic Energy Agency (IAEA), millions of radioactive sources (*radioaktivnie istochniki*) exist throughout the world. The IAEA estimates that thousands have either been lost (and potentially stolen) or are inadequately monitored and controlled.[28] This problem appears especially acute in the FSU.[29] IAEA records also indicate a sharp rise in reported incidents of smuggling of radiological materials.[30] The international community has made disposing of the approximately 1,000 radioisotope thermoelectric generators (RTGs) located in the FSU a priority. These remote power sources present both environmental and security hazards. Thieves and vandals often release radioactivity in the course of stripping precious metals from the sites, and terrorists could incorporate their radioactive materials into 'dirty bombs' (*gryaznie bombi*).

The Russian government has played a key role in securing the repatriation of Soviet/Russian-origin nuclear materials from Bulgaria, Georgia, Kazakstan, Libya, Romania, Serbia and other countries.[31] In fact, the success of these efforts, as well as the months of tedious interagency and multilateral negotiations required to implement them, convinced Moscow and Washington of the need for a more formal and comprehensive arrangement.[32] Under the GTRI's auspices, the Russian Research Reactor Fuel Return programme aims to repatriate civilian HEU fuel from Soviet/Russian-supplied research reactors. Rather than negotiate terms for each repatriation separately, the two governments signed an umbrella accord in May 2004 to govern all such operations. According to its terms, the United States will pay most of the costs incurred by Russian personnel in recovering and transporting the enriched uranium from the 17 countries that still possess Soviet/Russian-origin reactors for secure storage and possible recycling in Russia. Russian law also requires a complex environmental impact assessment before each shipment, and stipulates that some proceeds from the repatriation effort contribute to rehabilitating Russian territory contaminated by past nuclear activities.[33]

Both governments should consider making the GTRI more flexible by agreeing to permit the transfer of HEU from third parties to either country, not necessarily the state of origin. For example, if Iranian officials baulk at repatriating HEU from their US-supplied research reactor to the United States, GTRI funds should enable them to transfer the HEU to Russia, which enjoys far better relations with Tehran. Similarly, if a government objects to removing HEU from a Soviet-supplied reactor to Russia, GTRI managers should be able to finance its shipment to the United States. If feasible, the two governments might accelerate the entire repatriation effort, as recommended in 2004 by the UN High-Level Panel on Threats, Challenges and Change.[34]

It is clear that the GTRI will involve close cooperation between the United States, Russia and the IAEA. The IAEA often lacks the authority or resources to launch major non-proliferation initiatives on its own, but US and Russian assistance has enabled it to support GTRI projects. Russia and the United States could take further measures to strengthen the IAEA's contribution to global non-proliferation. For example, they could work to make its Model Additional Protocol a global standard for compliance with the Nuclear Non-proliferation Treaty (NPT). The protocol enhances the agency's inspection powers by providing its personnel with more information, broader access rights and the authority to use advanced detection technologies in the signatory country. Over 100 countries have signed the Additional Protocol, but fewer governments have ratified it, and several potential nuclear-weapon states remain outside its scope.[35] Russia and the United States should also consider pushing for a UN Security Council resolution that would require countries that withdraw from the NPT to return or destroy in a verifiable manner any nuclear-related imports they acquired while a party to the treaty.[36] Such a step might discourage countries from exploiting the so-called Article IV 'loophole', which allows member states to gain foreign help for their nuclear energy programmes, and then abruptly withdraw from the treaty and use the assistance to pursue nuclear weapons.

Another non-proliferation initiative Washington should consider more seriously are the recurring proposals that Russia import foreign spent nuclear fuel for long-term storage and reprocessing (*import na khraneniya i pererabotku inostrannoe otrabotannoe yadernoe toplivo*) at an International Spent Fuel Storage Facility under some kind of IAEA oversight. Zheleznogorsk, the most commonly suggested site, already contains Russia's largest nuclear-waste facility, which will need to be expanded and upgraded in any case to accommodate the country's growing domestic spent fuel stockpiles.[37] In June 2001, the Duma amended Russia's environmental legislation to permit the importation of spent nuclear fuel of foreign origin for 'temporary' storage, pending its reprocessing. The new law also allows any nuclear waste (*otrabotannye toplivnye sborki*) generated during reprocessing to remain in Russia, a practice at odds with that found in the few other countries that permit the reprocessing of foreign spent fuel. (Article 48 of the environmental law still bans the import of foreign radioactive waste.) Russia has already begun importing and storing small quantities of spent nuclear fuel from Soviet/Russian-origin research reactors in Bulgaria, Latvia, Libya, Romania, Serbia and Uzbekistan.[38] Relocating additional spent nuclear fuel to Russia would advance nuclear non-proliferation by depriving exporting countries of opportunities to reprocess it to extract plutonium. It would also remove fissile materials from places that often have worse safety and security procedures than Russia. According to the IAEA, developing countries account for 60% of the new nuclear reactors under construction.[39] The Russian government could devote some of the estimated $10–20bn in revenue it expects to earn from such imports to non-proliferation and threat-reduction projects in Russia and other countries.[40] Russia's nuclear industry might also use the spent nuclear fuel in the plutonium breeder reactors it hopes to build.[41]

Russia would require American approval for many of these transactions because some 80% of the world's non-Russian nuclear fuel originated in the United States, or has been irradiated in reactors of US origin.[42] Section 123 of the 1954 Atomic Energy Act requires that Congress approve any civil nuclear-cooperation agreement between the United States and another country. These accords typically stipulate that the recipient country cannot reprocess or alter the form or content of the US-supplied material without prior American consent. In fact, the US administration would need to negotiate, and Congress would need to approve, a separate agreement with Russia before it could import US-controlled spent nuclear fuel. Until now, American concerns about Russian–Iranian nuclear cooperation and Russian plans to reprocess the spent fuel into plutonium have – along with vocal environmental opposition and the limited capacity of Russia's storage and reprocessing facilities – stalled plans to import and store spent third-party fuel.[43] Evolving global non-proliferation concerns, and the expected increase in civilian nuclear-energy production resulting from rising oil and gas prices and the long-term effects of the Kyoto Protocol on non-renewable energy sources, could warrant a reassessment.[44] Russian officials have indicated that they want the IAEA to investigate how Russia and other countries could serve as guaranteed nuclear-fuel suppliers as part of the IAEA's 'multilateral nuclear approaches to the nuclear fuel cycle' initiative.[45] Bush and other US officials have expressed support for creating an IAEA-supervised international bank of nuclear fuel for countries that pledge to renounce developing their own enrichment or reprocessing facilities.

The February 2005 Bratislava declaration stated that Russia and the United States 'will jointly initiate security "best practices" consultations with other countries that have advanced nuclear programmes'. The two governments have already begun to share insights with other governments and the IAEA itself.[46] Non-governmental experts have offered ambitious proposals to apply more intrusive threat-reduction techniques to India, Iran, North Korea, Pakistan and other states.[47] Legal, financial, technical and political barriers would make it difficult for US government agencies to pursue most nuclear-security programmes in these countries. Most obviously, the proposed recipients would probably interpret American efforts to apply threat-reduction techniques as attempts to disrupt or spy on their most sensitive defence programmes. At worst, these governments could interpret US non-proliferation activities as strategies to acquire better targeting information for possible future military operations.[48]

Given these suspicions, these countries might be more willing to work with Russian personnel to enhance the safety and security of their WMD stocks. In some cases, Russian-only groups might make the most effective contribution. In other instances, Russian experts might usefully join IAEA-led assessment and assistance teams, reinforcing the agency's understaffed Office of Nuclear Security and related bodies. Russian specialists would be especially knowledgeable regarding Soviet-era equipment, but could easily consult with their Western and IAEA colleagues about foreign-origin technologies. Russian personnel could also exploit their contacts to discourage these governments from testing or transferring WMD or their related materials and technologies. Unlike American policymakers, Russian officials can proceed unencumbered by US legislative prohibitions against cooperating with states of proliferation concern. The optimal situation would be if the Russian government could provide funding and other resources for these endeavours, clearly demonstrating its status as a genuine partner in the pursuit of shared Russian and American security objectives.

Developing bilateral ballistic-missile defences

Russian and American officials have discussed possible cooperation in ballistic-missile defence (BMD) since the early 1990s. In fact, Russia currently deploys the world's only operational national missile defence (NMD) system, which it continues to upgrade.[49] In a January 1992 speech at the UN, Yeltsin called for a joint BMD research and development programme. The Clinton administration redirected the dialogue towards securing a bilateral 'demarcation' agreement that would permit theatre-wide missile defence (TMD) for forward-deployed military forces, while still preserving the prohibitions against large-scale NMD systems contained in the 1972 Anti-Ballistic Missile (ABM) Treaty.[50] When the Bush administration made clear its determination to proceed with NMD, Putin offered to relax the treaty's restrictions on testing. After this effort failed and US officials announced their unilateral withdrawal from the treaty in December 2001, Putin declined to make the issue a crisis. Instead, he and other Russian officials merely expressed regret for a decision they characterised as a 'mistake'.[51] They also announced that Russia no longer felt obliged to restructure its nuclear arsenal as required by the Strategic Arms Reduction Treaty (START I), especially its prohibition against placing more than one nuclear warhead on a single ballistic missile.

The US tactic of keeping Russian officials well-informed about America's BMD plans helped prepare the way for the even-tempered Russian response to the deployment decision. A stream of senior US representatives visited Moscow during 2001 in an attempt to find a formula by which Russia would accept a US national missile-defence system within an amended treaty. After this effort failed, both governments shared in advance the statements they planned to release to the public when announcing the US withdrawal from the ABM Treaty. Their coordinated, non-inflammatory response went far to dispel fears about a major deterioration in bilateral ties.[52] Moreover, by this time most Russian defence experts had concluded that the planned US BMD programmes remained too limited to place Russia's nuclear deterrent at risk. Putin himself acknowledged that they could not threaten Russia for several decades.[53] Russian analysts soon began to argue that Russian companies could therefore safely work with American entities to develop joint BMD technologies.[54] Russian aerospace, defence and other firms have long shown interest in such cooperation (and have persistently overestimated US interest in their potential contribution).[55]

The May 2002 Joint Declaration on the New Strategic Relationship between Russia and the United States, which sets out principles to govern their security cooperation, calls for bilateral measures to promote confidence, transparency and cooperation regarding BMD. Specifically, it advocates exchanging information on programmes and tests, reciprocal visits to observe BMD exercises, joint research and development on BMD technologies, and the implementation of a 1998 Yeltsin–Clinton agreement to create a Joint Data Exchange Center (JDEC) near Moscow to share early-warning information about missile launches worldwide, including possible accidental launches by their own countries.[56] In September 2002, the two governments formed a bilateral Missile Defence Working Group under the auspices of their Consultative Group for Strategic Security, established by the SORT to provide a forum for discussing and implementing bilateral BMD cooperation. A joint statement by Bush and Putin issued in June 2003 also endorsed accelerating BMD cooperation. At the time, the US Missile Defense Agency (MDA) expressed particular interest in gaining access to Russian expertise regarding radar systems and practice BMD targets.[57] Russian experts assert that, thanks to their proximity to potential rogue states (*stran-izgoev*), the 'ground-based radars of the Russian strategic early warning system possess unique capabilities to survey and control the missile threat directions in the vast area from the Middle East to the Korean peninsula – the main source of the threat for mankind today'.[58]

Despite these declarations, practical collaboration between Russia and the United States on missile defences has been minimal. Russian interlocutors typically depict the BMD issue as a distraction from the more pressing need to improve defences against terrorists employing other means of attack. They also fear that US BMD efforts will encourage China, India, Japan and other countries to acquire or increase their own nuclear arsenals.[59] Russian and US experts still have not completed negotiations on a bilateral military-technical framework agreement, which Russian officials have insisted they require to collaborate on BMD with the United States. Russia has signed such accords, which explicitly protect intellectual property and the confidentiality of Russian firms' proprietary information, with several other countries.[60] One way to overcome these differences might be to apply the kind of confidentiality agreements used by the joint working group supervising the DOE's installation of modern security systems at the Russian Navy's HEU storage facilities. These provisions require the parties' explicit consent before information shared within the group can be released into the public domain, unless it has been previously published elsewhere.[61]

Specific bilateral BMD projects suffer from additional problems. The development of the JDEC has been stymied by disagreements over liability, taxes and threat assessments.[62] Discussions that began in August 2004 about the possible use of Russian radars to help track North Korean missile launches continue without practical results.[63] MDA officials now express doubts about the benefits of developing a joint early-warning system or securing Russian rockets as targets for BMD interceptor missiles.[64]

Russian–US BMD exercises, which have occurred regularly since 1996, largely involve computer simulations (Command Post Exercises (CPX; *komandno-shtabnie ucheniya*, or KSHU)), rather than actual military units.

In discussing bilateral BMD cooperation, Putin has told journalists: 'We think that the time has not come to invest big money in such a project yet. We do not have this money to spare. We have good cooperation with our American partners and they are interested in working with our specialists. This is understandable. We do, after all, have some good developments in this area, developments that other countries, perhaps, do not have. We will work together, but we first need to agree on the principles for this cooperation'.[65] Defence Minister Ivanov has said that achieving concrete progress will take decades.[66] Ivanov has also specified several prerequisites for fruitful cooperation: (1) Russian and US BMD systems should not be directed at each other; (2) they should proceed with 'full transparency' (*polnaya prozrachnost'*) to both governments; (3) both sides' intellectual property must be protected; (4) outer space should remain demilitarised; and (5) bilateral BMD cooperation must take into account the national interests of Russia, the United States and, where necessary, third countries.[67] On a visit to Washington, General Yuri Baluyevsky, head of the Russian General Staff, added that such collaboration must be reciprocal: 'We are against collaboration that consists of "your ideas for our money". We favour cooperation based on joint ideas, joint resources, and shared results.'[68]

The two countries' experience with the Russian–American Observation Satellite (RAMOS) illustrates the pitfalls of BMD cooperation. RAMOS began in 1992 with the aim of jointly developing and operating two experimental missile-detecting satellites. Initially, they were to conduct remote sensing of the Earth's surface. By the end of the decade, it was hoped that the satellites would be able to track ballistic missiles. Although the two governments developed the requisite technologies, disputes over the ABM Treaty and other US BMD programmes impeded progress, and both sides accused the other of showing insufficient interest in the project. The Bush administration initially made a vigorous effort to revive RAMOS as part of its campaign to overcome Russian concerns about US NMD. MDA managers increasingly complained about its soaring costs, however, leading DOD to terminate the programme in February 2004.[69] Former MDA head Lieutenant-General Ronald Kadish once described RAMOS as the most realistic way for Russia and the United States to begin cooperating on BMD.[70] By 2004, however, Kadish had decided that RAMOS had been too ambitious, and that cooperation should focus on smaller projects.[71] One particular ambition of RAMOS – its extensive sharing of early-warning technologies and practices – might have made some participants especially uneasy.[72]

The currently restrained bilateral Russian–US dialogue on BMD, as well as the cautious multilateral collaboration occurring concurrently under NATO's auspices, shows that both parties now recognise the need for modest expectations in this controversial area. The only successful case of bilateral high-tech cooperation has been in the realm of civilian outer space, where the financial and other benefits (i.e., prestige) to Russia are much clearer. At present, neither Russian nor US officials anticipate such benefits in the case of BMD. Ivanov explicitly warned against any attempt to militarise space (such as by deploying space-based BMD interceptors or lasers).[73] Furthermore, Putin and other Russian political and military leaders highlight their plans to strengthen Russia's offensive nuclear forces so that they could overcome any conceivable defence.[74] American policy-makers profess unconcern, arguing that planned US BMD systems do not aim to counter Russia's missiles.[75]

Enhancing Russia's export controls

Moscow and Washington clashed over Russia's foreign military sales throughout the 1990s. US officials complained about the purchasers (such as India, Iran and other states of proliferation concern) and the items sold (especially ballistic missiles and their core components and technologies).[76] Russia's civilian nuclear-power cooperation with Iran has aroused particular ire. In January 1995, Russian officials agreed to help Iran to complete construction of a controversial $800m, 1,000-megawatt light-water reactor at Bushehr. Evidence later came to light that Minatom representatives had offered to assist Iran in developing its own uranium-enrichment capacity and other sensitive nuclear fuel cycle elements. US officials called on Russia to cease all nuclear collaboration with Iran because Tehran appeared to be seeking an indigenous nuclear-weapons capacity. Washington has made cooperation with Moscow in several high-technology sectors – including advanced civilian reactors and space exploration – conditional on Russia abandoning such work.

Russians increasingly appreciate US concerns about Tehran's nuclear ambitions, but they insist that their contribution to Iran's civilian programme adheres to NPT provisions and IAEA safeguards.[77] They stress that Russian negotiators secured Tehran's agreement in February 2005 to return all spent nuclear fuel from Bushehr to Russia, to preclude Iran from reprocessing it for weapons-usable plutonium (Bush called Russia's repatriation requirement 'constructive').[78] Moscow has also backed the efforts of the EU-3 (Britain,

France and Germany) to negotiate a non-proliferation deal with Tehran.[79] In the face of foreign criticism, Russian officials have accused Americans of employing a double standard. In particular, they point to the similarity between the two reactors the United States offered North Korea under the 1994 Agreed Framework (in return for its ceasing plutonium reprocessing) and the Bushehr plant. Claiming that European and even US companies cooperate with Iran on nuclear matters and were seeking to exclude Russian competitors, Putin said 'we object to … the use of the nuclear card as a means of bad-faith competition in the Iranian market'.[80]

Although Russia's nuclear commerce with Iran yields only a modest overall profit, it does provide benefits to influential Russian entities. In addition, Moscow hopes that the sales will lead Tehran to purchase other Russian products.[81] After the June 2005 presidential elections in Iran, Russian representatives reaffirmed their interest in constructing additional reactors.[82] Meanwhile, Iran's nuclear infrastructure and expertise may have progressed to a point where Russia's further involvement will have little impact on its ability to build an atomic bomb, particularly given the availability of alternative sources of black-market assistance.[83] Putting aside their past differences, Russian and US officials should evaluate whether the Russian–Iranian nuclear-fuel agreement might serve as a basis for establishing an international system of guaranteed nuclear-fuel services for countries that forgo indigenous enrichment and reprocessing capabilities.[84]

Although China, Pakistan and North Korea violate non-proliferation norms more egregiously, American officials continue to impose sanctions on Russian commercial entities suspected of selling WMD – especially missile-related elements – to problem states. A State Department report published in August 2005 found that, between January 2002 and December 2003, 'Russian entities continued to supply missile-applicable technologies to missile programmes of proliferation concern in China, India, Iran, and other countries'.[85] The relationship between the sanctioned firms and the Russian government, or particular Russian officials, is unclear. The State Department, itself uncertain about the connection, nevertheless observed: 'To the extent these transfers are not authorized by the Russian Government, they raise concerns regarding its ability to implement controls on missile-related technologies'.[86] Since most of the transactions at issue involve dual-use (*dvoynogo naznacheniya*) items suitable for both civilian and military purposes, many alleged violations invariably will not be clear-cut. In accordance with Executive Order 12938, the Iran Non-proliferation Act, the Missile Sanctions Law and other US legislation, typical penalties have included temporarily banning the companies concerned from US government contracts or from buying American defence equipment. These unilateral punishments have had limited practical effect since the targeted firms do not normally engage in such transactions, nor do they keep large assets in the United States, where they could be vulnerable to seizure.[87]

Thanks to several large barter deals and other creative arrangements, Russian arms deliveries have averaged $4–5bn in recent years, more than the government has spent on equipping its own armed forces.[88] Some of these transactions have disturbed US observers. For example, Russian offers to sell military helicopters and other weapons to Venezuela, and to supply Syria with anti-aircraft systems, have aroused unease in Congress and the executive branch. Russian government representatives consistently reply that their sales violate no laws, and that the United States and other countries also transfer large volumes of weapons to areas of conflict, including South Asia and the Middle East. They further contend that American protests often reflect a desire to eliminate unwelcome Russian competition, or to curtail Moscow's influence in important regions. In February 2005, Ivanov told the NRC: 'Export control regimes should not serve as a shield for undertaking bad-faith (*nedobrosovestnaya*) competition'. He also denounced 'double standards' and 'black lists'.[89] Russian officials criticise US sanctions policy as an extraterritorial application of national legislation to foreign enterprises in violation of international law. In addition, they point out that the US arms-export control system also suffers from imperfections – an assertion confirmed by the GAO.[90]

Disagreements over Russian arms exports will probably continue. In June 2005, the head of the advisory board to Rosoboronexport, the state arms-export agency, said that Russia planned to expand its exports to Latin America and Southeast Asia given the saturation of the Chinese and Indian markets.[91] Despite the visibility of military sales to Iran, Syria and other rogue states, China and India have purchased approximately 80% of Russia's recent arms exports.[92] Although Russia's defence sector has shrunk significantly since the Soviet era, it still suffers from limited domestic orders and extensive overcapacity. Since the Russian government cannot purchase sufficient weapons to keep its defence enterprises healthy, Russian officials have encouraged them to sell their wares abroad. In addition, the Russian government considers arms exports a vital source of foreign revenue and high-tech employment. Russian officials also appreciate that many defence companies require increased investment to develop the advanced systems that proved so effective for Western militaries in the Gulf, the former Yugoslavia and Afghanistan. Furthermore, they consider arms sales one of Moscow's best instruments for influencing the policies of foreign governments.[93]

Successive Russian governments have nonetheless taken steps to strengthen the legal and regulatory basis of the country's export controls for WMD-related items and missile technologies. In July 1999 the Duma enacted a comprehensive Federal Law on Export Control that defined the scope of the controls, the procedures for revising the lists of covered items, the government bodies charged with enforcement and the sanctions that would be imposed on individuals and commercial entities that violated the law's provisions. By the end of the Yeltsin era, however, Russia's export-control system still suffered from a variety of problems, including poor interagency coordination, widespread corruption, inadequate customs resources, weak enforcement of existing laws and leakage through quasi-independent territorial entities inside Russia (like Tatarstan), or through other former Soviet republics.[94]

The system has evolved further under Putin.[95] Presidential Edict No. 96, issued in January 2001, established the underlying framework of Russia's current export controls. In particular, it restructured the Federal Export Control Commission as an interagency body under the deputy prime minister for the military–industrial complex. At their June 2001 summit in Ljubljana, Putin and Bush formed a joint working group on export controls, and in August 2001 Putin issued Edict No. 1005, which identifies the missile-related equipment and technologies that are subject to regulation.[96] Russian authorities have also refined their procedures for controlling dangerous pathogens and BW-related dual-use items.[97] In March 2004, a major government restructuring created a Federal Technical and Export Control Service (FSTEK) within the Ministry of Defence. This body has the main administrative responsibility for implementing Russia's export controls.[98] On 26 April 2005, Putin signed Edict No. 468, which established a new export-control commission under the chairmanship of the defence minister. Its main functions include ensuring Russia's compliance with its arms-control obligations and, atypically, monitoring other countries' adherence to their own commitments to identify potential threats to Russia from WMD proliferation. The commission will also prepare annual reports on proliferation issues, and offer recommendations on how to improve international cooperation in this area.[99]

Russia and the United States cooperate in several multilateral institutions designed to curb WMD-related exports. Both countries are full participants in the Wassenaar Arrangement. This regime, established in April 1996, promotes transparency, responsibility and restraint in transfers of conventional arms and related dual-use products and technologies. They also adhere to the Missile Technology Control Regime (MTCR), which Russia formally joined in August 1995. The MTCR seeks to control the export of ballistic missiles and their underlying equipment or technologies to states of proliferation concern.

In April 2004, the UN Security Council unanimously adopted a joint Russian–US draft resolution that requires all countries to develop and maintain appropriate measures to (1) account for their WMD-related items in production, use, storage or transport; (2) safeguard them physically; (3) control their export and transshipment; and (4) inform their private industries and publics of their legal obligations regarding WMD proliferation. It also established a committee to monitor implementation. Since the resolution (UNSCR 1540) was adopted under Chapter VII of the UN Charter, it is binding on all members regardless of whether they formally adhere to international non-proliferation organisations, regimes and agreements. The resolution marked the first time that the UN Security Council has taken concrete measures to address the supply side of the global proliferation problem.[100] Thus far, however, flaws in implementation have weakened its impact. Russia, the United States and other governments need to work with the IAEA and the UN to specify the safety and export-control standards that would meet the resolution's requirement for 'appropriate effective border controls and law enforcement efforts'. They could also provide additional technical and financial assistance to countries that are experiencing difficulties in meeting these standards. In their summit declaration in June 2004, the G-8 leaders pledged such help.[101]

Cooperation between Russia and the United States against WMD proliferation broke new ground at the end of May 2004 when Russia, after a year of hesitation, formally agreed to join the US-led Proliferation Security Initiative (PSI; *Initsiativ po bezopasnocti v bor'be s rasprostraneniem* (IBOR)).[102] The PSI seeks to promote international agreements and partnerships to interdict possible shipments of WMD, their means of delivery (especially ballistic missiles) and their related equipment and technologies to and from 'states or non-state actors of proliferation concern'. Russia's participation in PSI activities has facilitated efforts to curb WMD proliferation across its huge land and air spaces. According to one US official, by late January 2005 Russian agents had twice attempted to intercept what intelligence analysts thought were shipments of missile components from North Korea to Iran, across Russian territory.[103] Russian officials themselves note their country's proximity to North Korea, Iran and other states of proliferation concern. In the words of Ivanov: 'The fight against WMD black markets pursued jointly with major foreign countries is fully consistent with Russia's best interests, since that real threat is geographically most close to us'.[104]

Despite these measures, a November 2004 report by the Central Intelligence Agency (CIA) stated that Russian entities were still trying to export technologies related to biological and chemical weapons and ballistic missiles to countries such as China, India and Iran. Although the CIA acknowledged that the Russian government had made 'progress in creating a legal and bureaucratic framework for Russia's export controls', the report complained that 'lax enforcement remains a serious concern'.[105] Ivanov has acknowledged that Russia still lacks an effective system for integrating non-proliferation and export controls.[106] Putin himself has decried the lack of coherence and trained specialists in Russia's export-control system.[107] US laws designed to influence Russian non-proliferation policies continue to limit bilateral cooperation in several high-technology sectors. For example, the 2000 Iran Nonproliferation Act makes collaboration between the US National Aeronautics and Space Agency (NASA) and Roscosmos, the Russian space agency, conditional on the president's certifying that Russia is not sharing nuclear or ballistic-missile technology with Iran. Despite Roscosmos' efforts to restrain the missile-exporting practices of Russian aerospace firms and other Russian entities, Bush has sometimes been unable to make the obligatory certification.[108]

Conclusion

Although they are making progress in reducing the technical barriers to greater interoperability, the Russian and US militaries still know little about each other. Continued bilateral dialogue and other exchanges are needed to overcome these problems. Russian and US military forces in Central Asia should develop direct ties, exchange information and perhaps conduct joint military exercises. Differences in threat perceptions, weapons technologies and financial resources make effective joint BMD collaboration unlikely. The two governments, however, are increasingly cooperating to limit the proliferation of WMD, their delivery systems and related materials. They are also exploring ways to pursue joint threat-reduction efforts in third countries through the Global Partnership and other institutions. In order to respond to foreign pressure, protect Russian technologies and other intellectual property and limit WMD proliferation, Moscow has strengthened its export controls. Nevertheless, differences over Russia's nuclear relationship with Iran in particular continue to impede potential Russian–American cooperation in several areas.

CHAPTER THREE

Strengthening NATO–Russian ties

Working through NATO provides Moscow and Washington with a useful multilateral supplement to their bilateral security cooperation. Formal ties between NATO and Russia developed soon after the end of the Cold War. In 1991, Russian representatives participated in the inaugural session of the North Atlantic Cooperation Council (NACC), the forerunner of the Euro-Atlantic Partnership Council (EAPC). These two institutions, along with the Partnership for Peace (PfP) programme, which Russia joined in June 1994, have provided mechanisms through which NATO countries and their former Soviet bloc adversaries could initiate defence and security cooperation. This first phase in the relationship culminated in the successful joint peacekeeping operation in Bosnia starting in December 1996; the May 1997 NATO–Russia Founding Act on Mutual Relations, Cooperation and Security; and the concurrent creation of the NATO–Russia Permanent Joint Council. During the late 1990s, relations were set back by the alliance's decision to offer several East European countries membership, and its military intervention in Kosovo, which had neither Russian nor UN approval. In response, the Russian government halted its participation in the PJC and suspended the operations of the NATO information office in Moscow.[1] Direct contacts only resumed after NATO Secretary-General George Robertson met Putin in Moscow in February 2000.[2] Subsequently, the 9/11 attacks vividly reminded Russian and Western officials that, on most security questions, their shared interests outweigh the issues that divide them.

Both parties took major steps towards the restoration of mutual cooperation with their joint declaration 'NATO–Russian Relations: A New Quality' in May 2002, and the simultaneous creation of the NATO–Russia Council. Consisting of all alliance members plus Russia, the NRC formally treats all members as equal partners, discarding the PJC's bilateral (and frequently confrontational) format. Operating on the basis of consensus, the NRC provides a mechanism for exchanging views, developing joint policies and programmes and, when possible, undertaking common or coordinated action. Russian representatives frequently cite these principles as the tenets that they would like other international institutions to follow. At a meeting of the Russian Security Council in January 2005, Putin affirmed the value Russia places on its cooperation with NATO: 'Our choice in favour of dialogue and cooperation with the North Atlantic Alliance has proven, in our opinion, to be correct and productive. It clearly has contributed to strengthening the international position of the Russian Federation. It has given us many new additional opportunities for solving national tasks'.[3] In April 2004, NATO and Russia signed agreements establishing Russian Military Liaison Offices at the NATO Operational Command in Mons, Belgium, and the NATO Transformation Command in Norfolk, Virginia. They also assigned additional personnel to the existing NATO Military Liaison Mission in Moscow.[4]

Russian interest in expanding security ties with NATO partly reflects an absence of institutional alternatives. Since ten additional countries, mostly former Soviet allies in Eastern Europe, entered the European Union (EU) in May 2004, Russian officials have increasingly complained about the EU's failure to offer Russia the same high level of security ties as NATO. Ivanov, for example, has lamented that 'cooperation between Russia and the EU in the defence and security sphere is still in the embryonic stage'.[5] Differences over trade, visa regimes, human rights, regional conflicts in former Soviet territories (especially Chechnya) and other issues have led Russian officials to attempt to bypass the EU and work directly with its most important members bilaterally, a strategy that has shown some success with France, Germany and Italy.[6] In addition, Russian interest in expanding the OSCE's security functions has declined markedly. During the 1990s, Moscow advocated making the OSCE Europe's predominant security institution.[7] In recent years, however, the organisation's efforts to promote human rights in Chechnya and encourage democratisation throughout the FSU have led Russian representatives to seek to cut its budget, limit its power and circumvent its activities.[8] Although Russian leaders advocate expanding the UN's security role, they recognise that the Bush administration remains wary of enhancing its authority.

Differences between NATO and Russia persist regarding implementation of the amended Conventional Forces in Europe (CFE) Treaty; the alliance's possible expansion further east (into Ukraine, for example); US nuclear weapons based in Europe; and military transparency in the airspace above the Baltic states. Moscow's current preoccupation is that NATO will establish military bases in the FSU. Despite these differences, Russians and Americans have cooperated within NATO on several important security areas that usefully supplement their direct bilateral ties. A distinct advantage NATO has over competing institutions is its greater experience in sharing sensitive information among its members. Effective cooperation between Russia and the West on terrorism, defence research and development and BMD typically requires the exchange of classified and proprietary data.[9] Opportunities for further, if limited, collaboration exist in these and other areas.

Countering terrorism

Cooperation against terrorism has become a priority for the NRC. Under its auspices, member countries have been conducting joint threat assessments of al-Qaeda in the Balkans, Islamic extremism in Central Asia and the terrorist threat to civil aviation.[10] In June 2004, NATO and Russian forces conducted a joint exercise, *Kaliningrad 2004*, to assess combined defences against terrorists using WMD. The scenario involved managing a mass-casualty environmental disaster caused by a terrorist attack. About 1,000 personnel from 22 NATO-affiliated countries and international organisations participated.[11] In December 2004, the NRC adopted a comprehensive action plan on terrorism that provides for increased cooperation to avert and manage terrorist attacks. The plan calls for sharing more intelligence, best practices and lessons learned from past incidents. It also envisages the joint development of new weapons and technologies aimed at responding to future terrorist threats.[12] Only half jokingly, Ivanov has suggested renaming NATO the 'New Anti-Terrorist Organisation'.[13] He has singled out the NATO–Russia conferences on the military's role in combating terrorism for generating 'lots of concrete and serious proposals'. He has also extolled the exchanges in intelligence data on terrorist threats among NRC members, and has called for establishing a NATO–Russian centre to detect and prevent terrorist acts.[14]

Russian firms appear eager to cooperate with NATO on developing anti-terrorism technologies.[15] Pooling Russian and NATO expertise and resources could prove especially useful in dealing with the threat of radiological dispersal devices. For example, these governments could work with private industry to limit the use of radioactive materials in agricultural, industrial, medical and scientific products. Terrorists could incorporate these materials into a conventional bomb to create a dispersal weapon that, while not as destructive as a true nuclear weapon or other WMD, could still sow panic and disrupt economic activity over whole metropolitan areas. As a preventive measure, Russian and NATO experts could assess the technical and economic feasibility of substituting non-radiological materials for radioactive agents in some products. Improvements in sensor technologies have gradually reduced the need for radioactive emitters, but security considerations might warrant accelerating this trend. Such a supply-side initiative would complement other national and international programmes, such as the Action Plan on Securing Radioactive Sources adopted by the G-8 governments at their summit in Evian, France, in 2003; the G-8 agreement in 2004 to limit their exports to legitimate end-users; and the Tripartite Initiative between Russia, the United States and the IAEA to track down and secure former Soviet radiological materials.[16]

In April 2004, the UN General Assembly adopted an International Convention for the Suppression of Acts of Nuclear Terrorism. This document, originally proposed by the Yeltsin administration, requires adhering countries to criminalise specific offences involving nuclear terrorism by non-state actors (e.g., using a radioactive device 'to cause death or serious bodily injury' or to damage property or the environment). Its other provisions seek to improve transnational information-sharing and law enforcement in this area. Russia's G-8 partners endorsed the convention at their July 2005 Gleneagles summit.[17] As primary potential targets of nuclear terrorism, Russia and the United States should take the lead in providing training, technology and finances to facilitate the convention's implementation.

Working closely with their NATO counterparts has encouraged the Russian defence community to recognise that terrorism encompasses more than just Chechnya. Differences over Russian tactics there significantly impeded such cooperation before 9/11.[18] Russia's decision to participate in NATO's *Operation Active Endeavour* will result in the two parties' armed forces operating together for the first time in a collective defence operation under Article 5 of the Washington Treaty.[19] NATO launched this maritime interdiction mission in October 2001 to counter international terrorism, WMD trafficking and other transnational threats in the Mediterranean. Although the Russian Navy plans to join NATO's Mediterranean patrols in 2006, Moscow's residual concerns about the alliance's expanding influence became evident when Russian officials dismissed proposals to extend *Active Endeavour* to the Black Sea, which many Russians still consider within their sphere of influence.

Ambitious proposals have been made to operationalise NATO–Russian cooperation further by establishing joint anti-terrorist units. Achieving such progress will require overcoming interoperability and other differences between NATO and Russia.[20] Russia's plans to establish a specially trained military brigade for multilateral operations might provide the nucleus of a joint anti-terrorist unit. In the interim, developing mechanisms to avoid friendly fire incidents in future counter-terrorist campaigns warrants increased attention. Russian representatives have begun mirroring US policies by affirming their right to launch pre-emptive strikes against terrorist groups.[21] One could easily imagine both NATO and Russian forces simultaneously attacking the same terrorist targets, which could prove disastrous if, for instance, US units had deployed surreptitiously in an area targeted by Russian missiles.

Consequence management

NATO and Russia first began working together on emergency planning and consequence management in December 1991, when the alliance helped to coordinate the delivery of humanitarian assistance to Russia following the Soviet Union's abrupt collapse. On 20 March 1996, following several years of joint workshops and other cooperation, NATO and Russia signed a Memorandum of Understanding on Civil Emergency Planning and Disaster Preparedness. The agreement committed the parties to develop a capacity for joint action in response to civil emergencies such as earthquakes and floods, as well as to cooperate to mitigate disasters.[22] In 1998, Russia helped to persuade other EAPC members to establish a Euro-Atlantic Disaster Response Coordination Centre to manage the consequences of natural and man-made emergencies. The centre maintains a database of national assets potentially available to assist with emergencies.[23]

Since the mid-1990s, Russian representatives have participated actively in civil emergency planning conducted under the auspices of the PfP. Russia has hosted a variety of exercises, seminars and workshops, especially after 9/11. At the May 2002 NATO–Russia summit, attendees identified consequence management as a priority for future cooperation. In September 2002, 14 NATO members and partner countries participated in *Bogorodsk 2002*, an exercise that simulated a terrorist attack on a chemical factory in Noginsk, Russia, and in June 2004 Russia hosted *Kaliningrad 2004*. Later that year, Russia and Hungary launched an initiative to strengthen the EAPC's ability to prevent, plan for and respond to civil emergencies. Meanwhile, the PfP Status of Forces Agreement (SOFA; *soglashenie o pravovoy statuse vooruzhennix cil po programme*), signed by NATO and Russia in April 2005, will, once it is ratified by the Duma, provide a legal framework to regulate the transit and deployment of military forces within each other's territories during consequence-management, peacekeeping and other joint exercises or operations. All of its provisions apply reciprocally: NATO troops in Russia have the same rights and privileges regarding documentation, taxation, customs and related issues as Russian troops in NATO territory.[24] The question of regulating Russian military transit to the Kaliningrad enclave across the territory of Lithuania (a NATO member) remains unresolved.

For almost a decade, the US Federal Emergency Management Agency (FEMA) has been working closely with Russia's Ministry for Civil Defence, Emergencies and the Elimination of Consequences of Natural Disasters (EMERCOM). On 16 July 1996, the heads of the two agencies signed a ten-year bilateral memorandum of understanding on cooperation. Since then, the two agencies have convened joint seminars and conferences and have organised scientific exchanges. They have also participated in each other's workshops, exercises and training courses. Then FEMA Director Michael Brown advocated developing joint response plans for major emergencies, and exploring combined assessment and response teams, with common command centres, in future humanitarian crises.[25] At their February 2005 Bratislava summit, Bush and Putin committed to improving their governments' capabilities for managing a nuclear/radiological incident. They also called for an international expert conference to derive lessons learned from the multinational response to the Indian Ocean tsunami of December 2004, and endorsed developing a bilateral coordination mechanism to improve their management of future humanitarian emergencies. These cooperative activities could easily encompass other EAPC members.

Finally, the Arctic Military Environmental Cooperation (AMEC) initiative has provided another cooperative framework between Russia and NATO countries. Under its auspices, the defence ministries of Russia, Norway, the United States and (since June 2003) the United Kingdom have addressed military-related environmental problems in the Arctic, especially the dismantling of Russia's nuclear submarines. Participants believe that cooperation among military rather than civilian agencies can best address these issues. Their close and productive interaction has helped to prevent the kind of access problems that have affected other Russian military sites.[26]

Several factors help to explain why Russian officials appear especially eager to cooperate with NATO in the area of consequence management.

Russians face numerous threats from both natural and man-made disasters, and they welcome foreign assistance in meeting these challenges. Second, collaborating on consequence management presents fewer interoperability or other problems than more challenging missions, such as joint combat operations. Finally, cooperation in this area is more balanced than is the case in many other arenas. Russia has niche assets, such as large transport aircraft and military units trained to operate in WMD-contaminated environments, which could assist NATO in a crisis.

Peacekeeping and post-conflict stability operations

The importance of NATO–Russian cooperation in peacekeeping and post-conflict stability operations ('*mirotvorcheskaya operatsiya*') becomes evident when one contrasts their experiences in Bosnia and Kosovo. In the former case, the hard-sought participation of Russian troops in the Implementation Force (IFOR), which deployed in December 1995, and in the follow-on Stabilisation Force (SFOR), established in December 1996, helped to secure a cease-fire between the combatants. It also eased Russian fears regarding the country's diminished international stature and declining role in European security affairs.[27] For several years, over 1,300 Russian soldiers conducted joint patrols, mine clearance, security checks and police actions in northern Bosnia. They also assisted with civil reconstruction and other tasks, such as the repatriation of refugees and displaced persons. The IFOR operation employed a creative command arrangement, in which the head of Russia's contingent reported as a Special Deputy to NATO's Supreme Allied Commander, Europe (SACEUR) in his capacity as Commander, US European Command. The commander of the Russian brigade in Bosnia came under the tactical control of the US general in charge of Multinational Division North.[28]

By contrast, the differences in early 1999 between NATO governments and Russia over how to respond to Serbian policies in Kosovo, combined with the alliance's decision to offer membership to Moscow's former East European allies, set back many NATO–Russian projects for years. NATO and Russian forces nearly came into direct conflict during the 'race to Pristina' in Kosovo in June 1999.[29] Although Russia subsequently provided the largest non-NATO contingent (some 1,500 troops) to the UN-mandated, NATO-led Kosovo Force (KFOR) in 2003, Moscow withdrew its forces from both Bosnia and Kosovo due to its lack of influence in these operations, NATO's perceived discrimination against ethnic Serbs and the belief that Russia could better use the troops elsewhere.[30]

Drawing on the Kosovo and Bosnia experiences, a working group set up by the NRC has developed a Generic Concept for joint peacekeeping operations. It establishes a framework for consultation, planning and decision-making during crises, and defines issues for combined training and exercises. After several years of meetings and other activities, NATO and Russian forces conducted a major exercise in September 2004 based on a draft Generic Concept. In early February 2005, the Russian government declared operational a special military unit, the 15th Detached Peacekeeping Motorised Rifle Brigade, whose main purpose is to work with NATO and other foreign militaries on international peacekeeping, search-and-rescue and counter-terrorist operations.[31] To enhance interoperability, its officers will learn English and other foreign languages, and will take classes at NATO schools. Foreign military personnel will receive similar training in Russia.[32] The fact that Russian policymakers have decided to bear the costs of training a 2,200-strong unit primarily for international peacekeeping operations testifies to the importance they assign this mission. The unit could help to fill the gap in practical NATO–Russian cooperation that has developed following the Russian military's withdrawal from Balkan peacekeeping missions.

The joint NATO–Russian operations in Bosnia and Kosovo exposed major differences in command structures, logistics, terminology and operating procedures. The experience also highlighted the divergence in their general approach to peacekeeping and especially peace enforcement. NATO and Russia have undertaken a long-term Interoperability Framework Programme to overcome these problems. This initiative has emphasised joint training and exercises, interoperability tests for equipment and procedures and other forms of concrete cooperation. In the belief that defence reform will enable the Russian military to work better with alliance forces, NATO and Russian policymakers have been sharing their experiences in this area. In July 2002, NATO established a joint programme for retraining retired Russian military personnel. The following year, the NATO Defence College established two fellowships for Russian analysts to research defence-reform issues. By 2004, NATO and Russia were engaged in approximately 40 annual events or projects related to defence reform.[33]

In October 2004, NRC defence ministers adopted a framework for Political–Military Guidance towards Enhanced Interoperability between Russian and NATO Forces.[34] This document will define the goals, tasks, procedures and other elements to govern joint NATO–Russian operations. In coordination with the Supreme Headquarters Allied Powers Europe (SHAPE), the NATO and Russian militaries presently engage in 45 specific projects aimed at improving interoperability. The cooperation plan for 2005 includes 13 joint exercises, seven meetings of operational staff officers, ten exercises designed to improve joint communications and ten meetings on military training.[35] To facilitate this elevated level of cooperation, Russia has established a permanent liaison office, led by a general officer, at SHAPE headquarters.[36] In October 2004, instructors from the NATO School in Oberammergau in Germany held courses on the alliance's structures, functions, operations and procedures at several Russian military academies.[37] Enhancing the parties' ability to communicate with one another – both technically (such as in equipment compatibility) and in terms of terminology and concepts (such as developing an agreed Concept of Operations for joint NATO–Russian peacekeeping operations) – remains a priority.[38]

NATO–Russian collaboration has made substantial progress in the maritime realm. The NRC has established several working groups on naval issues. In their 2002 Rome Declaration, alliance leaders highlighted search-and-rescue at sea as a subject for future cooperation. In February 2003, NATO and Russian officials signed a framework agreement to govern collaboration during submarine rescue operations. Subsequently, the Russian government initiated bilateral discussions with individual NATO countries, including the United States, on this issue. In 2004, the NATO Industrial Advisory Group invited Russian enterprises to work with Western defence firms to enhance the interoperability of submarine rescue systems. In 2005, a Russian ship was to participate in *Sorbet Royal*, a NATO-sponsored submarine search-and-rescue exercise. Unlike in August 2000, when the Russian government declined offers of foreign assistance in rescuing the disabled submarine *Kursk*, in August 2005 Moscow eagerly accepted Western help to rescue the crew of a stricken mini-submarine. Progress has also been made in other areas of maritime security. In 2004, Russian warships participated in 11 military exercises with NATO countries (three in the North Atlantic, four in the Baltic and four in the Mediterranean).[39] Since these were primarily bilateral, a logical next step would be to include Russian vessels in genuinely multinational exercises under NATO auspices. Although the presence of more than two navies would make such exercises more complex than their bilateral equivalents, they might expand the range of alliance missions in which Russian sailors could contribute.

NATO–Russian defence-industry collaboration

Increasing cooperation between NATO and Russian defence industries would further enhance military interoperability. During the 1990s, Russian experts and government officials began to attend NATO Advanced Research Workshops. In 1998, NATO and Russia signed a Memorandum of Understanding on Scientific Cooperation, and Russian scientists began to participate in the NATO Science Programme. Under the auspices of the NRC Science Committee, NATO and Russian experts have collaborated on several counter-terrorism projects, including detecting explosives and analysing the social and psychological effects of terrorism. They have also assessed military-related environmental problems within the NRC Committee on the Challenges of Modern Society. In 2002, the NRC established a working group for defence industry and technology. Among other steps, the group has arranged meetings where NATO and Russian firms can discuss possible collaboration. A forum on transport security in Moscow in April 2005 included representatives from EADS, Tales, Boeing, Finmekanika, Marconi Selenia Communications and Rheinmetal, as well as many Russian companies.[40] Much defence-industry cooperation has occurred under the auspices of NATO's Research and Technology Organisation (RTO), which sponsors approximately 150 defence-related research projects annually. Despite recent progress, several impediments continue to limit the involvement of Russian scientists, engineers and managers in RTO projects, including a lack of trust, inadequate Russian representation in technical forums, poor English skills and financial constraints on Russian participation in foreign events.[41]

Russian defence enterprises continue to repair, and even upgrade, some Soviet-era weapons (especially warplanes) in former Warsaw Pact countries. Nevertheless, Russian officials complain that they are losing business as new NATO members in Eastern Europe shift to purchasing equipment used by other allied militaries.[42] Some East Europeans suspect that Moscow wants to keep them dependent on Russian technology, and is seeking to gain access to military intelligence.[43] Russian firms have had even less success at selling products to long-established NATO allies. Russian companies might expand their sales further if they adopted more NATO Standardisation Agreements (STANAGs). These accords, which result from the deliberations of the Conference of National Armaments Directors (CNAD), establish processes, terms and conditions for common equipment and technical procedures between alliance members. The NATO Standardisation Agency in Brussels lists hundreds of STANAGs covering such things as the calibre of small-arms ammunition, map markings and communications procedures. For Russian companies to make greater use of NATO STANAGs, however, allied governments must agree to provide more advanced technological data. This will require Russian representatives to offer better guarantees regarding the non-transfer of such information to third parties, as well as overcoming seemingly excessive concerns about protecting military secrets and Russian intellectual property.[44]

Producing equipment more compatible with that found in NATO militaries would benefit Russia in two ways. First, its armed forces would be able to operate more effectively with NATO. Secondly, its defence companies could gain wider access to foreign armaments markets – though whether this opening would generate more sales depends on other factors, including marketing skills, relative competitiveness and government procurement policies. For their part, NATO countries could both collaborate better with Russian forces and obtain additional sources of defence equipment, possibly with superior technologies and at lower cost.

Ivanov has acknowledged that 'Interoperability is a difficult process and it will take many years to work it out. In this respect, we are only just embarking on this path'.[45] Nevertheless, he has expressed confidence that the superior quality of Russia's military systems will win over some NATO buyers.[46] In this regard, the barriers to achieving greater defence-industry cooperation and interoperability should not be exaggerated. Even within the alliance, member countries do not regularly acquire interchangeable equipment or technologies. Rather, the objective has been to achieve sufficient compatibility to enable allied forces to fight together with minimum friction and maximum effectiveness. This reflects a recognition that political and economic imperatives require members to manufacture some defence equipment themselves. The same considerations apply to NATO–Russian relations.

Cooperation in theatre-missile defence

Theatre-missile defence represents another area of possible cooperation between NATO and Russia. During a visit to Italy in June 2000, Putin indicated Moscow's interest in sharing TMD technologies with other European countries.[47] In February 2001, Russian officials proposed that they jointly develop with NATO a mobile land-based TMD system.[48] In June 2002, the NRC established an Ad-Hoc Working Group on Theatre Missile Defence (TMD AHWG). Its five Support Working Teams, composed of experts from NATO staff and member countries, have focused on terminology, experimental concepts, joint concepts of operations, systems and capabilities and training and exercises.[49] The parties began by defining common definitions and mutual threats, and are now assessing what types of compatible TMD systems could best meet these challenges.

In March 2004, the first NATO–Russian Command Post Exercise (CPX) occurred at the Joint National Integration Center in Colorado Springs, under the NRC's auspices. This computer simulation tested elements of the evolving TMD AHWG draft Concept of Operations (CONOPS), which would govern the use of TMD systems in a joint mission outside alliance territory.[50] Another CPX, *Collaborative Arrow '05*, was held in March 2005 at De Peel airbase in the Netherlands. This exercise involved a ten-day test of the experimental CONOPS, practicing in particular procedures for joint NATO–Russia TMD planning and operations. Russia has offered to host the next TMD CPX in the second half of 2006.[51]

One factor driving Russian interest in TMD collaboration is the hope that NATO countries will purchase Russian missile-defence technologies and weapons systems. Deputy Foreign Minister Sergei Kislyak said in 2003 that Russians 'have our own anti-missile systems that might be useful, and they are among the world's best … we are very serious partners'.[52] Ivanov has offered to contribute the S-300 and forthcoming S-400 air-defence systems to a future European TMD system, including one directed against the growing threat of cruise missiles.[53] Russian representatives claim that, unlike their own firms, American defence suppliers refuse to transfer their best BMD technologies, insist on manufacturing most items domestically and expect allies to pay dearly for any purchases.[54]

Compared with other NATO–Russian issues, progress on joint TMD has been slow. This partly reflects the difficulties and financial costs associated with deploying any missile-defence system. The fact that the two sides have been developing systems that employ different technical standards, command-and-control procedures and operational engagement doctrines has also impeded progress, as have genuine disagreements over the ballistic-missile threat. Most Russian defence analysts discount the danger from ballistic missiles relative to other security challenges. NATO governments had been cooperating for many years on BMD projects before they decided, primarily for policy reasons, to incorporate a Russian contribution. The longstanding ties between NATO defence firms, and their superior information technologies, have limited their interest in collaborating with Russian companies. Restrictive technology-transfer policies impede cooperation among close NATO allies; the barriers between alliance members and Russia are even greater. Russian defence firms also lag behind NATO companies, for instance in terms of optimal size and competitiveness.[55]

Russian officials have evinced alarm about NATO's decision to upgrade US early-warning radars in Greenland and the UK, and about possible plans to station American BMD assets in Eastern Europe. Ivanov has said that these locations imply that Russia remains the primary target of US BMD efforts. He told an Italian newspaper in February 2005 that 'If such a decision were to be made in earnest, it could seriously weigh down the activities conducted in the context of the NATO–Russia Council on anti-missile weapons and it could negatively impact the entire Euro-Atlantic security system'.[56] Some analysts have characterised vague Russian offers to cooperate on TMD as efforts to dissuade Europeans from supporting the US NMD system.[57]

NATO's decision in March 2005 to develop an Active Layered Theatre Ballistic Missile Defence (ALTBMD) system by the end of the decade might spur renewed NATO–Russian cooperation in this area. The ALTBMD will provide a common integrating structure for individual allied TMD systems, to improve their collective ability to protect deployed troops. NATO governments are also investigating defences against long-range ballistic-missile strikes against their territories and populations.[58] On the one hand, Russian military and business leaders might more eagerly participate in an actual acquisition programme, as opposed to a joint concept-development process. On the other hand, they are likely to oppose an ALTBMD that excluded Russia. NATO's TMD plans appear to have prompted a Russian inquiry in 2005 regarding how Washington would react if Moscow withdrew from the 1987 Intermediate-Range Nuclear Forces (INF) Treaty (*Dogovor o likvidatsii raket sredney i men'shey dal'nosti*). This pioneering accord banned all Soviet and US ground-launched ballistic and cruise missiles with ranges between 500 and 5,000km.[59] Such missiles could help Russia to overcome a NATO TMD system. Clearly, the record of past TMD collaboration affirms the need for modest expectations. Rather than extensive sharing of technology or establishing a common TMD architecture, a more attainable goal would be to enhance interoperability and achieve a better understanding of each party's TMD development plans and CONOPS.

Dealing with tactical nuclear weapons

One potential stumbling-block to NATO–Russian cooperation in general concerns the two parties' stockpiles of tactical nuclear weapons (TNW; *takticheskoe yadernoe oruzhie* (TYAO)). Although the 1991 and 1992 Presidential Nuclear Initiatives (PNI) eliminated many types of TNW, and removed other systems from operational deployment, NATO analysts argue that the Russian government should consider bringing its remaining arsenal into the CTR programme (which has thus far only encompassed strategic warheads).[60] Advocates of TNW arms control fear that the small size of these weapons, and their scattered location, mobility, and weaker security and safety features, make them more vulnerable to terrorist seizure than strategic warheads.[61] Members of the US Congress have repeatedly urged the executive branch to make a greater effort to secure Russia's TNW.[62] Secretary of Defense Donald Rumsfeld told a Senate hearing in July 2002 that he had raised the TNW issue in every meeting with his

Russian interlocutors, but without result.[63] In early June 2005, Assistant Secretary of State for Arms Control Stephen Rademaker said that Russian officials continued to evince 'very little interest in talking to us'. A few days later, the State Department complained that Moscow had failed to provide adequate information regarding the fulfilment of its PNI commitments.[64]

Russian officials have resisted extending threat-reduction activities to their TNW partly because they believe that their opacity contributes to deterring a pre-emptive NATO attack. Uncertainties regarding the number and location of Russia's tactical nuclear weapons mean that NATO planners cannot be sure of destroying them in a first strike. Such considerations weigh against proposals to consolidate Russia's TNW, even if dispersal makes them more vulnerable to terrorists. Russian analysts also note that TNW represent one of the few areas where Russia enjoys military superiority over NATO.[65] Securing Moscow's agreement to consolidate and better secure these weapons may require concessions from the US regarding its tactical nuclear weapons based in Europe. Russian leaders, who point out that all of their TNW now lie solely within Russian territory, have complained about the continued deployment of these weapons. Although Russian concerns about a NATO military attack have declined, Baluyevsky, the head of the Russian General Staff, observed in late 2003 that the hundreds of US air-deliverable TNW in Europe 'are for Russia acquiring a strategic nature since theoretically they could be used on our command centers and strategic nuclear centers'.[66] In early June 2005, Ivanov said that Russia was 'prepared to start talks about tactical nuclear weapons only when all countries possessing them keep these weapons on their own territory'.[67] Rademaker termed these remarks a 'stalling tactic' to prevent negotiations, observing that 'it is a very convenient position for the Russians to take because they can withdraw their tactical nuclear weapons to Kaliningrad ... and say that they have withdrawn [their TNW] to national territory and why doesn't the United States do the same'.[68] Even the redeployment of all US TNW in Europe to North America might prove insufficient. Russian officials note that Washington could return them in a few hours unless NATO irreversibly destroyed their storage sites and related infrastructure.[69] Verifying any agreement could prove difficult given that their delivery systems (e.g., attack aircraft) are typically dual-use systems that can also launch conventional strikes.

For these reasons, technical assistance programmes to enhance the safety and accounting of Russia's TNW might prove less difficult to implement, especially if they focused on rear-area storage facilities, rather than forward-located operational sites.[70] A first step would be for both governments to share information regarding the number, condition, and security and safety procedures of their TNW. If future conditions permit, they could consolidate or dismantle additional systems. According to the Russian media, Russia will soon start a ten-year programme to upgrade thousands of its existing tactical nuclear weapons with a smaller number of next-generation systems, so any safety and security programme should begin in the near future.[71]

Conclusion

NATO has undergone a sweeping transformation since the end of the Cold War. It has admitted seven new members, reworked its strategy and conducted its first actual military operations. It has also expanded ties with Russia in several important areas. NATO and Russia have shared intelligence and threat assessments, conducted joint exercises to improve their response to terrorist attacks and natural disasters and established a legal basis to govern joint operations. NATO and Russian defence planners have reduced some interoperability problems, but overcoming non-technical barriers will require further work. The two parties also differ over TMD and especially TNW. Despite their disagreements, Russians appreciate that NATO will remain Europe's pre-eminent military institution, and that the alliance will continue to provide opportunities for mutually profitable security collaboration.

CONCLUSION

Towards Better Russia–US Security Relations

This survey of Russian–American security cooperation suggests several overarching conclusions. On the one hand, market-based incentives and reciprocity seem to promote bilateral ties. On the other, disputes over Russian–Iranian nuclear collaboration and the disillusionment following unmet expectations represent two major short-term impediments to revitalising security cooperation between Russia and the United States.

Using more market incentives

The Iranian case shows the power of market incentives to shape non-proliferation behaviour. The US policy of imposing sanctions on Russian firms that contribute to Iran's nuclear-weapon or ballistic-missile programmes has primarily influenced companies that receive US-funded threat-reduction or space-cooperation contracts. These firms naturally do not want to jeopardise these deals. The same holds true within the Russian government. Roscosmos, which cooperates closely with NASA on civilian space issues, has more readily embraced US non-proliferation practices than Minatom/Rosatom, which sees exporting nuclear fuel and technology as vital to its institutional survival. During the 1990s, the Clinton administration effectively presented the Russian government with a choice: either sell rocket engines to India, or participate in the International Space Station as a subsidised partner. Moscow chose the latter course.[1] Similarly, Russian companies' (misplaced) expectations that they could sell BMD technologies to NATO members has stimulated Russian interest in cooperating with Western countries on BMD. In the case of Iran, however, the Clinton and Bush administrations have been unable to convince most Russian leaders or businesses that working on Tehran's nuclear programmes could substantially damage their ties with Washington.

Exploiting additional market-based incentives could also improve bilateral threat-reduction programmes. One reason for the relatively effortless enforcement of the HEU–LEU Agreement is that Russia receives billions of dollars from the contract. In contrast, the present lack of demand for Russia's excess weapons-grade plutonium has reduced incentives for the timely implementation of the Plutonium Disposition Agreement. In many threat-reduction programmes, employing Russian-made equipment and Russian firms has reduced costs, built local constituencies for threat-reduction activities and transferred business skills to Russian entrepreneurs.

Promoting reciprocity and equal treatment

Russian representatives have repeatedly made clear their desire to be treated more as partners, and less as clients. They have requested additional financial and other data concerning US-funded threat-reduction projects in Russia, and have insisted that progress on BMD requires equal opportunities for Russian businesses, full transparency and mutual respect for national interests. Moscow also demands more equitable liability provisions in bilateral threat-reduction agreements. Russia's reluctance to cease cooperation with Iran's nuclear programme or end exports of advanced conventional weapons to China, Venezuela and other countries partly reflects a belief that Western countries conduct similar transactions. Cooperation with NATO became truly institutionalised only after the creation of the NRC which, unlike the PCJ, formally treated Russia as an equal. Bush has acknowledged the importance of permitting more Russian visits to US WMD-related sites in order to gain greater American access to Russian facilities.

Moscow and Washington could do much more than exchange additional data and visits. In return for increased financial and other contributions, Russia's role in designing and operating threat-reduction activities could expand. As more equal partners, Russia and the United States could conduct parallel or even joint audits of programmes, a process that would help expose and overcome misperceptions. The two governments could jointly lead international non-proliferation efforts within the Global Partnership, the UN and the GTRI. Relaxing visa requirements or strengthening controls on TNW will require reciprocal compromises. Opportunities for joint peace-keeping, counter-terrorism and other military operations will increase the more Russia's armed forces become interoperable with NATO.

Getting over Iran

US legislation prohibits various forms of cooperation with Russia in retaliation for its involvement in Iran's civil nuclear programme. Moscow and Tehran insist that their collaboration adheres to international law and practice. In July 2005, they initiated discussions on how Russia might help to construct up to 20 additional nuclear-power plants.[2] A programme of such dimensions could disrupt Russian–US relations for years. Russian–Iranian nuclear ties have been a major factor preventing Washington's endorsement of Russia's possible construction of an International Spent Fuel Storage Facility. They have also prevented US scientists from participating in joint research and development projects, which could result in more proliferation-resistant technologies and fuel cycles. In addition, the US Congress has repeatedly withheld other assistance to Russia because of Moscow's nuclear and ballistic-missile cooperation with Tehran.[3]

The rationale for continuing these negative policies has become weaker over time. Iran's nuclear infrastructure and expertise have progressed to a point where Tehran may have the capacity to manufacture nuclear weapons even without additional Russian help. Most Russians oppose Iran's acquisition of a nuclear arsenal, but they want to earn revenue, and generally downplay the risks of Tehran's actually attaining an atomic bomb. Although Iran's nuclear ambitions are worrying, opposition to an International Spent Fuel Storage Facility risks impeding broader US non-proliferation goals. Consolidating spent nuclear fuel in one or several secure locations would end the current worldwide proliferation of temporary storage sites, which multiplies opportunities for diversion and misuse. Furthermore, the Russian government could earn an estimated $10–20bn for storing and reprocessing the imported spent fuel. Moscow could devote some of this money to promoting non-proliferation, both within and outside Russia. Punitive US legislation also limits Russia's potential contribution to civilian nuclear-power research and the exploitation of outer space, which is particularly disturbing given the problems with NASA's shuttle fleet. A more productive approach would seek to apply to other countries the provision of the Russian–Iranian agreement requiring the return of Russia's nuclear fuel after its use at Bushehr. The United States should also applaud more vocally Moscow's efforts to block suspected shipments of missile components from North Korea to Iran across Russian territory. Russia's actions highlight the potential for greater bilateral and multilateral cooperation on PSI and related activities aimed at disrupting WMD-related proliferation. Preventing Iran from becoming a nuclear-weapon state is important, but restricting mutually beneficial cooperation with Russia is a poor way to proceed.

The need for modest expectations

Perhaps the most important conclusion is the need for modest expectations regarding Russian–US security cooperation. Setbacks in Kosovo, BMD and other areas demonstrate that rapid progress is unlikely. Although the Russian government is financing more threat-reduction projects, Russian resources will remain limited. Competitive economic and security pressures loom large in their bilateral relations, and Russian officials remain sensitive to US and NATO military activities in regions they consider within Moscow's sphere of influence.

When the Soviet Union first collapsed, some Americans and many Russians thought that the new Russian Federation could become another strategic partner of the United States, like Britain. British policymakers consider a strong America essential to their security, and actively promote US initiatives throughout the world. Similarly, Americans anticipated that Russians would come to support a strong US global presence. Many Russians initially hoped that, in return for backing US policies in Europe and elsewhere, Americans would shower them with aid and investment, and treat Moscow as a privileged partner in world affairs. Russia's unanticipated weaknesses, combined with Western stinginess, dashed these expectations.

After mutual disillusionment set in around 1993, Russian and US policymakers came to see something like the Franco-American relationship as a possible model. Unlike their British counterparts, French officials do not believe that American power invariably promotes their strategic interests, but France's limited resources, and the core values both nations share, circumscribe their rivalry. Although French policymakers routinely criticise US policies, they do not normally try to thwart them. Influential Russian and American leaders likewise have assumed that Russia's need for at least minimal US economic support, and Washington's desire for Moscow's help in fighting terrorism, curbing WMD proliferation and managing regional conflicts, require that the two countries work together despite their differences.

Putin's perceived authoritarian policies and disputes over Kosovo, Iraq and Iran have raised the spectre of a third model. Some Russians have drawn parallels between the troubled Sino-American relationship and Moscow's problematic ties with Washington. Influential figures in Russia and China have called for joint measures to counter American power, and the two countries have conducted an unprecedented series of joint military exercises. For their part, many Americans identify Russia and China as the most likely rivals to US interests in Eurasia.

The relationship between Russia and America will not approximate that between the United States and Britain, but ties can certainly be better than those between Washington and Beijing. Notwithstanding setbacks, Russians enjoy greater civil liberties than Chinese citizens, and still aspire to closer ties with the United States and other Western countries. Polls show that Americans and Russians genuinely desire better relations, and that Russians overwhelmingly favour allying with the United States rather than China.[4] This study has identified many opportunities for mutually profitable cooperation. French and American policymakers often distrust each other, but they recognise that their core interests overlap more than they differ. Similarly, Russian–American relations should reflect an appreciation that the two countries can best promote their security through cooperation, rather than confrontation.

Acknowledgements

I would like to thank Amira Ali, Nicole Aronzon, Hans Binnendijk, Stephen Blank, Caitlin Brand, Eric Brewer, Chris Brown, Sebastian Elischer, J. Charles Griggs, Adrianne Grunblatt, Edeanna Johnson, Ty Matsdorf, Caroline Patton, Klementina Sula, Noemi Szekely, Samir Tata, Brian Wender, Nick Wetzler, Krystal Wilson, Elizabeth Zolotukhina and several anonymous reviewers for their comments on earlier drafts of this paper. The Project on Nuclear Issues at the Center for Strategic and International Studies, run by Senior Fellow Clark Murdoch and Project Coordinator Kathleen McInnis, organised several conferences that allowed me to deliver presentations on Russian–American security issues. The Hudson Institute's Washington Office, directed by Ken Weinstein, and its Center for Future Security Strategies, directed by S. Enders Wimbush, provided an exceptionally favourable environment for conducting research and writing. Many Russian and US policymakers and experts shared their insights with me on an off-the-record basis. Finally, I benefited enormously from the seminars, presentations and discussions organised by the other think-tanks in the Washington area, whose participants are too numerous to list, but whose ideas continue to drive progress in this vital area.

Notes

Introduction

1   *Freedom in the World-2005* (New York: Freedom House, 2005), p. 519.

2   *Final Report on the Presidential Election in the Russian Federation, 14 March 2004* (Vienna: Organisation for Security and Cooperation in Europe), p. 1, at http://www.osce.org/documents/odihr/2004/06/3033_en.pdf. See also Anders Aslund, 'Putin's Decline and America's Response', *Carnegie Endowment for International Peace Policy Brief*, no. 41, August 2004, pp. 1–2.

3   See for example Stephen Dinan and Jeffrey Sparshott, 'Senators Seek To Sanction Russia', *Washington Times*, 18 February 2005.

4   See James M. Goldgeier and Michael McFaul, 'What To Do About Russia', *Policy Review*, no. 133, October–November 2005, http://www.policyreview.org/oct05/goldgeier.html.

5   The transparency problem is discussed in Alexei G. Arbatov, 'Military Reform: From Crisis to Stagnation', in Steven E. Miller and Dmitri V. Trenin (eds), *The Russian Military: Power and Policy* (Cambridge, MA: MIT Press, 2004), pp. 97–98. See also the figures in Michael McFaul, 'Reengaging Russia: A New Agenda', *Current History*, vol. 103, no. 675, October 2004, p. 308. According to surveys of Russian business

leaders conducted by the Russian Indem Foundation, Russian business people pay over $300bn in bribes annually. Dmitri Trenin, 'Reading Russia Right', *Carnegie Endowment Policy Brief*, no. 42, October 2005, p. 3. The widespread prevalence of corruption in Russia's economy is also discussed in Steven Myers, 'Pervasive Corruption in Russia Is "Just Called Business"', *New York Times*, 13 August 2005; and 'Blood Money', *The Economist*, 22–28 October, 2005.

6 See for example Pavel Felgenhauer, 'New Détente To Die Young', *Moscow Times*, 29 May 2005; Nikolai Sokov, 'Russian Ministry of Defense's New Policy Paper: The Nuclear Angle', *CNS Reports*, at http://cns.miis.edu/pubs/reports/sok1003.htm; and 'Russia Cites US Action for War Exercises', *International Herald Tribune*, 11 February 2004.

7 The Jackson–Vanik Amendment, contained in Title IV of the 1974 Trade Act, applies to certain countries, including Russia and several other former Soviet republics, with non-market economies that restrict emigration rights. Specifically, it denies them unconditional or permanent normal trade relations with the US unless the president determines that they comply with the amendment's emigration requirements. US presidents have found Russia in compliance with the amendment every year since 1994, but concerns about growing anti-Semitism in the FSU and other malign developments have prevented its repeal despite repeated Russian complaints about its humiliating biannual reviews.

8 The present and planned future nuclear forces of Russia and the United States are described in Richard Weitz, 'Resurgent Russia Confronts US Ally in Europe', in *The Future Security Environment and the Role of US Nuclear Weapons in the 21st Century* (Washington DC: Center for Strategic and International Studies, 2005), pp. 66–76. The *Bulletin of the Atomic Scientists* regularly publishes updates on Russian and US nuclear-weapon programmes and polices; see for example Robert S. Norris and Hans M. Kristensen, 'NRDC; Nuclear Notebook: Russian Nuclear Forces, 2005', in the March–April 2005 edition, vol. 61, no. 2, pp. 70–72; and Robert S. Norris and Hans M. Kristensen, 'NRDC; Nuclear Notebook: US Nuclear Forces, 2005', in the January–February 2005 edition, vol. 61, no. 1, pp. 73–75. The Arms Control Association also produces periodic fact sheets on both countries' strategic nuclear forces at http://www.armscontrol.org/factsheets/sovforces.asp and http://www.armscontrol.org/factsheets/usstrat.asp. The Nuclear Threat Initiative puts detailed information about Russia's nuclear stockpiles and other security issues on its website, at http://www.nti.org/db/nisprofs/russia/tc_ru.htm.

9 Keith Payne, 'The Nuclear Posture Review: Setting the Record Straight', *Washington Quarterly*, vol. 28, no. 3, Summer 2005, pp. 143, 146–148.

10 Although classified, excerpts of the NPR have been posted on the Internet at http://www.globalsecurity.org/wmd/library/policy/dod/npr.htm.

11 See for example Statement of Douglas J. Feith, Senate Armed Services Committee, Hearing on the Results of the Nuclear Posture Review, 14 February 2002, at http://www.globalsecurity.org/military/library/congress/2002_hr/feith0214.pdf; and DOD, *Annual Report to the President and Congress: 2002* (Washington DC: US Government Printing Office, 2002), p. 89.

12 For example DOD, *Annual Report to the President and Congress* (Washington DC: US Government Printing Office, 1997), p. 11, described Russia as a potential threat 'not because its intentions are hostile, but because it controls the only nuclear arsenal that can physically threaten the survivability of US nuclear forces'. See also Ivan Oelrich, *Missions for Nuclear Weapons after the Cold War* (Washington DC: Federation of American Scientists, 2005), pp. 14, 41, 44–45, 54.

13 This debate is reviewed in Nikolai Sokov, 'Modernization of Strategic Nuclear Weapons in Russia: The Emerging New Posture', May 1998, at http://www.nti.org/db/nisprofs/over/modern.htm; and Frank Umbach, *Future Military Reform: Russia's Nuclear & Conventional Forces* (Camberley: Conflict Studies Research Centre, Defence Academy of the United Kingdom, August 2002), pp. 11–14.

14 Yuriy Maslyukov, 'Pravo na otvetniy udar: Rossiya dolzhna soxranit' sistemu yadernogo cderzhivaniya', *Voenno-Promishlenniy Kur'er*, 22–29 October 2003; and Dmitri Trenin, *Russia's Nuclear Policy in the 21ˢᵗ Century Environment* (Paris: IFRI, Autumn 2005), pp. 11–12. See also Brad Roberts, *Trilateral Stability: The Future of Nuclear Relations Among the United States, Russia, and China* (Alexandria, VA: Institute for Defense Analysis, 2002), pp. 15, 33.

15 The government expects to spend almost \$340m on nuclear weapons in 2006; see ITAR-Tass, 17 August 2005, in *Global Security Newswire*, 18 August 2005. Additional information on Russia's nuclear weapons and war plans can be found in David Holley, 'Russia Seeks Safety in Nuclear Arms', *Los Angeles Times*, 6 December 2004; Robert S. Norris and Hans M. Kristensen, 'NRDC Nuclear Notebook: Russian Nuclear Forces, 2004', *Bulletin of the Atomic Scientists*, vol. 60, no. 4, July–August 2004, pp. 72–74; and Paul Webster, 'Just Like Old Times', *Bulletin of the Atomic Scientists*, vol. 59, no. 4, July–August 2003, pp. 30–35.

16 For a history of Russian–US nuclear nonproliferation collaboration, see Jim Walsh, *Russian and American Nonproliferation Policy: Success, Failure, and the Role of Cooperation*, MTA Occasional Paper 2004-01 (Cambridge, MA: Kennedy School of Government, Harvard University, June 2004). Russia's commitment to nonproliferation is analysed in Dmitri Trenin, 'Russia and Global Security Norms', *Washington Quarterly*, vol. 27, no. 2, Spring 2004, pp. 63–77.

17 Andrew Kuchins, Vyacheslav Nikonov and Dmitri Trenin, *US–Russian Relations: The Case for an Upgrade* (Moscow: Carnegie Endowment for International Peace, 2005), p. 7. For a history of Russia's counter-terrorism policies see Dmitri Trenin, 'Russia and Anti-Terrorism', in Dov Lynch (ed.), *What Russia Sees*, Chaillot Paper 74 (Paris: Institute for Security Studies of the European Union, January 2005), pp. 99–114.

18 The latter incident is discussed in GAO, *Nuclear Nonproliferation: US and International Assistance Efforts to Control Sealed Radioactive Sources Need Strengthening* (Washington DC: May 2003), p. 70.

19 These contrasting perspectives are discussed in Frances G. Burwell, *Re-Engaging Russia: The Case for a Joint US–EU Effort* (Washington DC: The Atlantic Council, February 2005), esp. p. 2.

20 For a discussion of the initial frictions between the Bush and Putin administrations over Chechnya and other issues, and Putin's surprising decision to support the US military response after the 9/11 attacks, see Peter Baker and Susan Glasser, *Kremlin Rising: Vladimir Putin's Russia and the End of Revolution* (New York: Scribner, 2005), pp. 121–123, 129, 132; George Friedman, *America's Secret War: Inside the Hidden Worldwide Struggle between America and Its Enemies* (New York: Doubleday, 2004), pp. 141–149; and Andrew S. Weiss, 'Russia: The Accidental Alliance', in Daniel Benjamin (ed.), *America and the World in the Age of Terror: A New Landscape in International Relations* (Washington DC: Center for Strategic and International Studies, 2005), pp. 125–134. The Churchill analogy appears in Baker and Glasser, *Kremlin Rising*, p. 135.

21 These joint exchanges between Russian and US executive-branch agencies are discussed in DOS, 'Europe and Eurasia Overview', *Country Reports on Terrorism 2004*, 27 April 2005, at http://www.state.gov/s/ct/rls/45388.htm.

22 The World Bank, *Russian Economic Report* (Washington DC: February 2004), p. 3.

23 DOS, 'Background Note: Russia', February 2005, at http://www.state.gov/r/pa/ei/bgn/3183.htm.

24 See for example the figures in Michael McFaul, 'Reengaging Russia: A New Agenda', *Current History*, vol. 103, no. 675, October 2004, p. 308.

25 For a discussion of some of these projects, see 'Has the Russian Space Launch Quota Achieved Its Purpose?', in Hearing before the International Security, Proliferation, and Federal Services Subcommittee of the Committee on Governmental Affairs, US Senate, 19 July 1999.

26 See for example Alexei Arbatov, 'Superseding US–Russian Nuclear Deterrence', *Arms Control Today*, vol. 35, no. 1, January–February 2005, p. 14; George Perkovich, 'Bush's Nuclear Revolution: A Regime Change in Nonproliferation', *Foreign Affairs*, vol. 82, no. 2, March–April 2003, p. 8; and Jon B. Wolfsthal and Tom Z. Collina, 'Nuclear Terrorism and Warhead Control in Russia', *Survival*, vol. 44, no. 2, Summer 2002, pp. 71–83.

27 Andrey Poshshilin, *RIA Novosti*, 3 May 2005, reprinted in *Yaderniy Kontrol': Informatsiya*, 29 April–5 May 2005, at http://www.pircenter/org/data/publications/yki9-2005.html.

28 Cited in Wade Boese, 'US–Russian Nuclear Rivalry Lingers', *Arms Control Today*, vol. 35, no. 1, January–February 2005, p. 43.

29 Wade Boese, 'US Sets Missile Defense for Europe, Space', *Arms Control Today*, vol. 35, no. 4, May 2005, p. 30.

30 For a review of possible operational arms-control measures involving Russia and the United States see Oleg Bukharin and James Doyle, 'Transparency and Predictability Measures for US and Russian Strategic Arms Reductions', *Nonproliferation Review*, vol. 9, no. 2, Summer 2002, pp. 1–19; Lawrence Korb and Peter Ogden, *The Road to Nuclear Security* (Washington DC: Fourth Freedom Forum, December 2004); David E. Mosher et al., *Beyond the Nuclear Shadow: A Phased Approach for Improving Nuclear Safety and US–Russian Relations* (Santa Monica, CA: RAND, 2003); and Sergei Rogov et al., 'Russia and the United States: Reducing Mutual Nuclear Risks', briefing prepared for the Nuclear Threat Initiative (Moscow: Institute of the USA and Canada of the Russian Academy of Sciences, 2004), at http://www.ceip.org/files/pdf/Rogov05-20-04presentation.pdf; or as a report, Sergei Rogov et al., *Reducing Nuclear Tensions: How Russia and the United States Can Go Beyond Mutual Assured Destruction*, 19 January 2005, at http://www.nti.org/c_press/analysis_mad_011905.pdf.

Chapter One

1 State Department, *United States Initiatives to Prevent Proliferation*, 2 May 2005, at http://www.state.gov/t/np/rls/other/45456.htm. About $7bn of this total has been spent on enhancing the security of Russia's nuclear weapons; see Carla Anne Robbins and Alan Cullison, 'In Russia, Securing Its Nuclear Arsenal is an Uphill Battle', *Wall Street Journal*, 26 September 2005. The relative financial importance of these programmes is evident in the State Department chart, 'US Assistance to Russia – Fiscal Year 2005', 1 June 2005, at http://www.state.gov/p/eur/rls/fs/48458.htm, and the corresponding fact sheets for earlier years.

2 Strictly speaking, the term Cooperative Threat Reduction only applies to DOD programmes, but analysts often group all US government threat-reduction activities in Russia under the 'CTR' label.

3   National Intelligence Council, *Annual Report to Congress on the Safety and Security of Russian Nuclear Facilities and Military Forces* (Langley, VA: Central Intelligence Agency, December 2004), at http://www.cia.gov/nic/special_russianuke04.html). See also CIA Director Porter Goss's remarks concerning missing Russian nuclear material during his 16 February 2005 testimony before the Senate Select Committee on Intelligence, at http://intelligence.senate.gov/0502hrg/050216/goss.pdf.

4   'Russia Plans Security Exercises at Nuclear Sites', *Global Security Newswire*, 3 July 2004.

5   Charles L. Thornton, 'The G8 Global Partnership against the Spread of Weapons and Materials of Mass Destruction', *Nonproliferation Review*, vol. 9, no. 3, Autumn/Winter 2002, p. 143.

6   Rose Gottemoeller, 'Nuclear Weapons in Current Russian Policy', in Miller and Trenin (eds), *The Russian Military*, p. 209.

7   The Secretary of Energy Advisory Board, *A Report Card on the Department of Energy's Nonproliferation Programs with Russia* (Washington DC: DOE, 10 January 2001), at http://www.seab.energy.gov/publications/rpt.pdf.

8   *The 9-11 Commission Report: Final Report of the National Commission on Terrorist Attacks Upon the United States* (Washington DC: US Government Printing Office, 2004), p. 381.

9   Office of the Democratic Whip, *Ensuring America's Strength and Security: A Democratic National Security Strategy for the 21st Century*, September 2005, p. 5, at http://democraticwhip.house.gov/docuploads/nationalsecdofinal.pdf.

10   James Kitfield, 'Reducing the Threat', *National Journal*, 30 April 2005, p. 1,311.

11   Bryan Bender, 'Cut in Funds for Securing Nuclear Materials Rejected', *Boston Globe*, 7 January 2005. The White House overturned the reduction. William Hoehn has written helpful summaries of the DOD, DOE and DOS FY2006 threat-reduction budget requests at http://www.ransac.org/Publications/ Congress%20and%20Budget/Federal%20 Budget%20and%20Congressional%20Up dates/index.asp.

12   For more on these incidents see Justin Bernier, 'The Death of Disarmament in Russia?', *Parameters*, vol. 34, no. 2, Summer 2004, pp. 90–91; Office of the DOD Inspector-General, *Cooperative Threat Reduction: Cooperative Threat Reduction Program Liquid Propellant Disposition Project*, no. D-2002-154, Washington DC, September 2002; Office of the DOD Inspector General, *Cooperative Threat Reduction: Solid Rocket Motor Disposition Facility*, no. D-2003-131, Washington DC, September 2003; and GAO, *Cooperative Threat Reduction: DOD Has Improved Its Management and Internal Controls, But Challenges Remain*, Washington DC, June 2005, Appendix I, pp. 25–27. See also *Cooperative Threat Reduction Annual Report to Congress: Fiscal Year 2006* (Washington DC: DOD, 2005), pp. 4–10.

13   Igor Kudrik et al., *The Russian Nuclear Industry: The Need for Reform*, Bellona Report, vol. 4 (Oslo: Bellona Foundation, 2004), pp. 115, 164.

14   See for example the testimony of Ambassador Linton Brooks, then acting administrator of the NNSA, at the Hearings of the House Armed Services Committee, 4 March 2003; the testimony of Paul Longsworth, NNSA deputy administrator, at the Hearings of the Oversight and Investigations Subcommittee of the House Energy and Commerce Committee, 24 May 2005; and Joel Wit and Ian Woodcroft, 'United States', in Robert J. Einhorn and Michele A. Flournoy (eds), *Protecting Against the Spread of Nuclear, Biological, and Chemical Weapons: An Action Agenda for the Global Partnership*, vol. 3 (Washington DC: Center for Strategic and International Studies, January 2003), p. 242.

15   Michael Nguyen, 'Russia Expects To Double Funding for Chemical Weapons Disposal Activities', *Arms Control Today News Update*, 7 September 2004, at

http://www.armscontrol.org/aca/mid-month/2004/September/Russia.asp.

16 *Nuclear Nonproliferation: DOE's Efforts to Assist Weapons Scientists in Russia's Nuclear Cities Face Challenges* (Washington DC: GAO, 3 May 2001).

17 William Hoehn, 'Update on Congressional Activity Affecting International Threat Reduction and Cooperative Nonproliferation Programs', RANSAC Policy Update, August 2005, p. 21, at http://www.ransac.org/Publications/Congress%20and%20Budget/Federal%20Budget%20and%20Congressional%20Updates/index.asp.

18 US National Academies Committee on US–Russian Cooperation on Nuclear Non-proliferation and Russian Academy of Sciences Committee on US–Russian Cooperation on Nuclear Non-proliferation, Development, Security and Cooperation, National Research Council, *Overcoming Impediments to US–Russian Cooperation on Nuclear Nonproliferation: Report of a Joint Workshop* (Washington DC: National Academies Press, 2004), p. 44.

19 'US–Russia: No Verification Measures Planned for Moscow Treaty', *Global Security Newswire*, 24 July 2003.

20 For a detailed description of these localities, see Yuriy Rumyantsev and Aleksey Kholodov, 'Conversion Challenges in Russian Nuclear Cities', *Nonproliferation Review*, vol. 10, no. 3, Autumn/Winter 2003, pp. 167–182.

21 *Weapons of Mass Destruction: Additional Russian Cooperation Needed To Facilitate US Efforts to Improve Security at Russian Sites* (Washington DC: GAO, March 2003), p. 7.

22 GAO, *Cooperative Threat Reduction: DOD Has Improved*, pp. 5, 19.

23 For an assessment of the effects of Russian regional and local actors on threat-reduction issues, see Adam N. Stulberg, 'Nuclear Regionalism in Russia: Decentralization and Control in the Nuclear Complex', *Nonproliferation Review*, vol. 9, no. 3, Autumn/Winter 2002, pp. 31–46.

24 Nabi Abdullaev, 'A Bush Deal and a Missing Paragraph', *Moscow Times*, 1 March 2005; and Pavel Felgenhauer, 'Stymied by Nuclear Secrecy', *ibid.*, 26 April 2005.

25 For a history of the Trilateral Agreement on Biological Weapons of 10 September 1992, see Joseph Cirincione, Jon B. Wolfsthal and Miriam Rajkumar, *Deadly Arsenals: Nuclear, Biological, and Chemical Threats*, second edition (Washington DC: Carnegie Endowment for International Peace, 2005), p. 61; and Michael Moodie, 'The Soviet Union, Russia, and the Biological and Toxin Weapons Convention', *Nonproliferation Review*, vol. 8, no. 1, Spring 2001, pp. 62–63.

26 See, for example, Paula A. DeSutter, Testimony before the House Armed Services Committee, 4 March 2003, at http://www.state.gov/t/vc/rls/rm/18736.htm; State Department, *Adherence to and Compliance with Arms Control, Nonproliferation and Disarmament Agreements and Commitments*, Washington DC, August 2005, pp. 27–31; and Kenneth N. Luongo et al., 'Building a Forward Line of Defense: Securing Former Soviet Biological Weapons', *Arms Control Today*, vol. 34, no. 6, July–August 2004, pp. 18–23.

27 Jeff Zeleny, 'Strains from Biological Weapons Program Sent to US', *Chicago Tribune*, 2 September 2005.

28 Cited in Simon Saradzhyan, 'FSB Says Terrorists Are Trying To Secure WMD', *Moscow Times*, 22 August 2005.

29 Hoehn, 'Update on Congressional Activity', pp. 15–16.

30 GAO, *Cooperative Threat Reduction: DOD Has Improved*, p. 19.

31 DOS, *Adherence and Compliance*, p. 61.

32 *Ibid.*, pp. 59, 60.

33 DOD, *Cooperative Threat Reduction Annual Report to Congress: Fiscal Year 2005*, Washington DC, 2004, p. 4.

34 Matthew Bunn and Anthony Wier, *Securing the Bomb 2005: The New Global Imperatives* (Cambridge, MA: Kennedy

School of Government, May 2005), pp. 28, 31.

35 Kenneth Luongo and William Hoehn, 'An Ounce of Prevention', *Bulletin of the Atomic Scientists*, vol. 61, no. 2, March–April 2005, p. 30. A 'trusted agent' in this context would be a Russian individual or legal entity who enjoyed the trust of the American government and access to sensitive Russian sites, and who could convincingly verify the implementation of threat-reduction projects.

36 David Holley, 'US–Russian Efforts To Protect Arsenal Gain Steam', *Los Angeles Times*, 27 August 2005.

37 'Russia Continues To Resist US Access to Nuclear Sites Despite Security Cooperation, NNSA Chief Says', *Global Security Newswire*, 3 October 2005.

38 'Russia Not To Allow Observers To Visit Nuclear Sites – DM', *MosNews*, 9 August 2004, at http://www.mosnews. com/news/2004/08/09/ivanov.shtml. For descriptions of *Avaria 2004*, see Aleksandr Emel'Yanenkov, 'Avariya' na pyat' s plyusom', *Rossiyskaya Gazeta*, 7 August 2004; Sergey Severinov, '"Krash-test" dlya Yadernyx Boezaryadov', *Krasnaya Zvezda*, 17 August 2004; and 'Nuclear Weapons Accident Response Exercise Held in Murmansk Region', *NATO Update*, 25 August 2004, at http://www.nato.int/docu/update/2004/08-august/e0803a.htm.

39 Office of the White House Press Secretary, 'President Holds Press Conference', 20 December 2004, at http://www.white-house.gov/news/releases/2004/12/200412 20-3.html.

40 Bunn and Wier, *Securing the Bomb 2005*, p. 28

41 The 12th Main Directorate is abbreviated as GUMO in Russian.

42 Interview with chief of the Ministry of Defence's 12th Main Directorate Igor Valinkin, *Izvestiya*, 24 May 2005.

43 'US, Russia Plan to Defend Nuclear Installations', *Agence France-Presse*, 12 July 2005, at http://www.spacewar.com/news/nuclear-blackmarket-05zi.html.

44 Rose Gottemoeller, 'Arms Control in a New Era', *Washington Quarterly*, vol. 25, no. 2, Spring 2002, pp. 53–54.

45 GAO, *Cooperative Threat Reduction: DOD Has Improved*, pp. 5, 16–17; and *Overcoming Impediments to US–Russian Cooperation*, pp. 7, 10, 34, 94–95.

46 Additional data on the types and numbers of employees (technicians as well as weapon scientists) at each Russian WMD facility, each site's future expansion or downsizing plans and the activities of their retirees and other former employees would be especially helpful; see John V. Parachini et al., *Diversion of Nuclear, Biological, and Chemical Weapons Expertise from the Former Soviet Union: Understanding an Evolving Problem* (Santa Monica, CA: RAND Corporation, 2005), p. 43.

47 'Brickbat: Visa Rub', *Albuquerque Tribune*, 5 April 2004, at http://web.abqtrib.com/archives/opinions04/050504_opinions_edwed.shtml; and James W. Brosnan, 'Guests of Labs See Visa Trouble', *Scripps Howard News Service*, 5 April 2004, at http://web.abqtrib.com/archives/news04/050404_news_visa.shtml.

48 *Border Security: Streamlined Visa Mantis Program Has Lowered Burden on Foreign Science Students and Scholars, but Further Refinements Needed* (Washington DC: GAO, February 2005).

49 *Nuclear Nonproliferation: Security of Russia's Nuclear Material Improving; Further Enhancements Needed* (Washington DC: GAO, February 2001), p. 15.

50 'Russia Reducing Foreign Inspections at Chemical Sites, Increasing Technical Monitoring, Official Says', *Global Security Newswire*, 12 November 2004.

51 US Department of State, 'Midpoint of the Successful Implementation of the Highly Enriched Uranium Agreement Between the United States and Russia', Press Statement, 30 September 2005, at http://www.state.gov/r/pa/prs/ps/2005/54146.htm.

52 Julian Evans, 'Alarm Over Radioactive Waste Site', *Moscow Times*, 15 July 2005.

53   Matthew Bunn, Anthony Wier and John P. Holdren, *Controlling Nuclear Warheads and Materials: A Report Card and Action Plan* (Cambridge, MA: Kennedy School of Government, March 2003), pp. 76–77.

54   Charles Digges and Igor Kudrik of the Bellona Foundation describe the programme as the Russian nuclear industry's 'lifeblood' ('US Funding in Russia Should Encourage Nuclear Reform in Moscow', *Bellona Position Paper*, 3 June 2005), at http://www.bellona.no/en/international/russia/nuke_industry/co-operation/38312.html. See also Oana C. Diaconu and Michael T. Maloney, 'Russian Commercial Nuclear Initiatives and US Nuclear Nonproliferation Interests', *Nonproliferation Review*, vol. 10, no. 1, Spring 2003, pp. 104–106.

55   Gunnar Arbman et al., *Eliminating Stockpiles of Highly Enriched Uranium: Options for an Action Agenda in Co-operation with the Russian Federation*, Swedish Nuclear Power Inspectorate, April 2004.

56   Peter Slevin, 'US–Russia Plutonium Disposal Project Languishing', *Washington Post*, 10 May 2004; and Matthew Wald, 'US–Russian Plan to Destroy Atom-Arms Plutonium is Delayed', *New York Times*, 9 February 2004.

57   Unlike conventional nuclear fuel, MOX is a blend containing about 5% plutonium oxide and 95% uranium oxide. The process of irradiating the plutonium agents in a commercial nuclear reactor makes them unusable for weapons production. Experts disagree, however, whether the resulting spent nuclear fuel will be sufficiently radioactive to discourage terrorist use, and about the security and safety risks of transporting and manufacturing MOX fuel. The arguments for and against this technique are summarised in Charles Digges, 'Technical Agreement for Plutonium Disposition Allowed To Lapse by US', *Bellona Foundation*, 30 July 2003, at http://www.bellona.no/en/international/russia/navy/co-operation/30596.html.

58   For a summary of the shutdown agreement and a detailed assessment of its potential problems, see *Nuclear Nonproliferation: DOE's Effort To Close Russia's Plutonium Production Reactors Faces Challenges, and Final Shutdown Is Uncertain* (Washington DC: GAO, June 2004).

59   For a description of the different immobilisation procedures and other issues related to MOX, see Charles Digges, 'MOX Eludes Mention at Evian G-8 Summit', 13 June 2003, at http://www.bellona.no/en/international/russia/nuke_industry/co-operation/29844.html.

60   Siegfried S. Hecker, 'Thoughts about an Integrated Strategy for Nuclear Cooperation with Russia', *Nonproliferation Review*, vol. 8, no. 2, Summer 2001, pp. 10, 16. See also Kudrik et al., *The Russian Nuclear Industry*, pp. 160–161. For signs that some influential Americans may be changing their views, see Steve Fetter and Frank N. von Hippel, 'Is US Reprocessing Worth the Risk?', *Arms Control Today*, vol. 35, no. 7, September 2005, pp. 6–12.

61   Bunn and Wier, *Securing the Bomb 2005*, p. 71.

62   David Albright and Kimberly Kramer, 'Tracking Plutonium Inventories', *ISIS Plutonium Watch*, August 2005, p. 2.

63   Einhorn and Flournoy (eds), *Protecting Against the Spread of Nuclear, Biological, and Chemical Weapons*, vol. 1, pp. 13–15, 32.

64   GAO, *Security of Russia's Nuclear Material Improving*, p. 18.

65   'US, Russia Resolve Some Issues Plaguing Chem Demil Program', *Inside the Pentagon*, 23 September 2004.

66   Igor Khrippunov, 'Export Control Assistance to Russia and Other FSU States', in *Protecting Against the Spread of Nuclear, Biological, and Chemical Weapons*, vol. 2, pp. 152–53.

67   Parachini et al., *Diversion of Nuclear, Biological, and Chemical Weapons Expertise from the Former Soviet Union*. Survey data suggest that foreign grant programmes have established relationships between Russian and Western

weapon scientists that have helped discourage the former from working with
rogue states and non-state actors; see
Deborah Yarsike Ball and Theodore P.
Gerber, 'Russian Scientists and Rogue
States: Does Western Assistance Reduce
the Proliferation Threat?', *International
Security*, vol. 29, no. 4, Spring 2005, pp.
50–77, esp. pp. 72–74, 76–77.

68 Steve Goldstein, 'Experts: Program to
Secure Enriched Uranium "Slow"',
Philadelphia Inquirer, 9 February 2004.

69 Peter Scoblic, 'United States, Russia
Approve New "HEU Deal" Contract',
Arms Control Today, vol. 32, no. 6, July–
August 2002, p. 20.

70 Possible ways to accelerate the purchase
and elimination of Russia's HEU stocks
are discussed in George Perkovich et al.,
*Universal Compliance: A Strategy for Nuclear
Security* (Washington DC: Carnegie
Endowment for International Peace,
March 2005), pp. 105–107; and Robert L.
Civiak, 'Closing the Gaps: Securing High
Enriched Uranium in the Former Soviet
Union' (Washington DC: Federation of
American Scientists, May 2002), at http://
www.fas.org/ssp/docs/020500-heu/. See
also the options developed by a joint
Russian–US expert group established
at the May 2002 Bush–Putin summit, as
described in 'Joint Statement: Secretary
Abraham and Minister Rumyantsev', 16
September 2002, at http://www.ne.doe.
gov/home/09-16-02.html. The Nuclear
Threat Initiative is funding a Russian
study to assess various options for
increasing the amount of Russian HEU
blended down annually; see *NTI Annual
Report 2004*, Washington DC, 2004, p. 18.

71 For a summary of the ISTC's activities in
Russia, see its 2004 *Annual Report* at http://
www.istc.ru.istc/website.nsf/html/04/en/
index.htm.

72 Jeffrey Read, 'Reported Accomplishments
of Selected Threat Reduction and
Nonproliferation Programs, By Agency,
for Fiscal Year 2004', *RANSAC Policy
Update*, July 2005, p. 6, at http://www.

ransac.org/Publications/Reports%20and
%20Publications/Reports/index.asp.

73 These problems are reviewed in Derek
Averre, Kenneth N. Luongo and
Maurizio Martellini (eds), *Advancing
Bio Threat Reduction: Findings From an
International Conference* (Washington DC:
Russian–American Nuclear Security
Advisory Council, 2004); Maurizio
Martellini and Kenneth Luongo, 'The G-8
Global Partnership Initiative: Prioritizing
Nonproliferation and Security Concerns',
5 May 2003, unpublished paper at http://
www.sgpproject.org/events/LuongoMart
elliniPaper050503.htm; and *Reshaping US–
Russian Threat Reduction: New Approaches
for the Second Decade* (Washington DC:
Carnegie Endowment for International
Peace and Russian–American Nuclear
Security Advisory Council, 2002), pp.
34–35.

74 On the more general problems with
US-funded threat-reduction efforts to
promote commercially viable defence
conversion in the FSU, see Henry D.
Sokolski and Thomas Riisager (eds),
*Beyond Nunn–Lugar: Curbing the Next
Wave of Weapons Proliferation Threats from
Russia* (Washington DC: Nonproliferation
Policy Education Center, April 2002).

75 Kenneth N. Luongo et al., 'Security Culture
in the NIS', *University of Georgia Center for
International Trade and Security Monitor*, vol.
11, no. 1, Spring 2005, pp. 5–6.

76 John Mintz and Joby Warrick, 'US
Unprepared Despite Progress, Experts
Say', *Washington Post*, 8 November 2004.

77 Sue Vorenberg, 'Siberian Challenges
Beckon to Bioscientists', *Albuquerque
Tribune*, 10 November 2003.

78 Derek Averre, 'From Co-option to
Cooperation: Reducing the Threat of
Biological Agents and Weapons', in
Einhorn and Flournoy (eds), *Protecting
Against the Spread of Nuclear, Biological, and
Chemical Weapons*, vol. 2, p. 45.

79 Russian military activities in this area
are discussed in 'Tsena protivoyadiya ot
bioterrorizma' [interview with Lt-Gen.

Vladimir Filippov], *Krasnaya Zvezda*, 13 November 2004.

80 Additional areas of possible collaboration between Russian and US weapon scientists are described in *Strategies for Russian Nuclear Complex Downsizing and Redirection: Options for New Directions* (Washington DC: Russian–American Nuclear Security Advisory Council, June 2003), pp. 3, 25.

81 'Observations on US Threat Reduction and Nonproliferation Programs in Russia', Statement of Joseph A. Christoff, Director, International Affairs and Trade, GAO, Before the House Armed Services Committee, Washington DC, 5 March 2003, pp. 1–2, 5–6.

82 For a discussion of Russia's persistent economic weaknesses, see Eugene B. Rumer and Celeste A. Wallander, 'Russia: Power in Weakness?', *Washington Quarterly*, vol. 27, no. 1, 2003, pp. 57–73.

83 See, for example, Einhorn and Flournoy (eds), *Protecting Against the Spread of Nuclear, Biological, and Chemical Weapon*, vol. 1, pp. 3, 12–13; and Nancy Pelosi and Harry Reid, 'How to Effectively Confront Nuclear Threat from Terrorists', *USA Today*, 25 August 2005.

84 Bunn and Wier, *Securing the Bomb 2005*, pp. 32–34, 67–68; and National Intelligence Council, *Annual Report to Congress on the Safety and Security of Russian Nuclear Facilities and Military Forces*, February 2002, at http://www.cia.gov/nic/special_russiannnucfac.html.

85 Zhanna Voronova, *RIA Novosti*, 17 June 2005, reprinted in *Yaderniy Kontrol': Informatsiya*, 15–22 June 2005, at http://www.pircenter/org/data/publications/yki16-2005.html.

86 Michael Nguyen, 'Russia Speeds Chemical Weapons Disposal', *Arms Control Today*, vol. 35, no. 1, January–February 2005, p. 44.

87 *ITAR-TASS*, 30 June 2005, reprinted in *Yaderniy Kontrol': Informatsiya*, 29 June–6 July 2005, at http://www.pircenter/org/data/publications/yki18-2005.html.

88 'Russia No Longer Funding Bioterrorism Countermeasures Research, Scientist Says', *Global Security Newswire*, 2 February 2005.

89 Jenifer Mackby and Ola Dahlman, 'Bioterrorism and a Layered Approach to Biodefense', *SGP Issue Brief*, no. 5 (October 2005), at http://www.sgpproject.org/publications/publications_index.html#SGPIssueBriefs.

90 Ivan Lebedev, *ITAR-TASS*, 17 June 2005, reprinted in *Yaderniy Kontrol': Informatsiya*, 15–22 June 2005, at http://www.pircenter/org/data/publications/yki16-2005.html.

91 C. J. Chivers, 'Keeping That Special Glow Safe at Home', *New York Times*, 2 May 2005.

92 For more on the taxation issue, see *Overcoming Impediments to US–Russian Cooperation*, pp. 28–29, 79–81, 95. Specific recommendations for changing the Russian tax code appear in US National Academies Committee on US and Russian Cooperative Nuclear Nonproliferation and Russian Academy of Sciences Committee on US and Russian Cooperative Nuclear Nonproliferation, *Strengthening US–Russian Cooperation on Nuclear Nonproliferation* (Washington DC: National Academies Press, 2005), pp. 25–26.

93 Paul F. Walker of Global Green USA, presentation at RANSAC 2005 Congressional Seminar Series, 15 July 2005.

94 The CTR Umbrella Agreement's formal name is 'Agreement Between the United States of America and the Russian Federation Concerning the Safe and Secure Transportation, Storage, and Destruction of Weapons and the Prevention of Weapons Proliferation'. For a discussion of the conditions pertaining to its provisional extension in 1999, see Susan Koch, 'Cooperative Threat Reduction Negotiations: Lessons Learned', in *Strengthening US–Russian Cooperation on Nuclear Nonproliferation*, pp. 61–68.

95 Peter Eisler, 'Renewal of Deal To Help Secure Russian Arms in Doubt', *USA Today*, 14 December 2004.

96 See, for example, Rumyantsev's comments in an interview in Andrey Zlobin, 'Na chustvetel'nye yadernie ob'ekti Rossii dostup amerikantsam zakritt', *Vremya Novostey*, 12 May 2005, reprinted in *Yaderniy Kontrol': Informatsiya*, 5–12 May 2005, at http://www.pircenter/org/data/publications/yki10-2005.html.

97 Keith J. Costa, 'US Trumpets Progress with Russia on Plutonium Disposition Liability', *Inside the Pentagon*, 12 May 2005.

98 Christine Kucia, 'Liability Concerns Jeopardize Renewal of Nonproliferation Programs With Russia', *Arms Control Today*, vol. 33, no. 7, September 2003, p. 40. The NCI has since been incorporated into DOE's new Global Initiatives for Proliferation Prevention programme.

99 See for example the remarks of John Bolton, cited in Mike Nartker, 'Bolton Takes Heat for Plutonium Disposal Effort', *Global Security Newswire*, 16 June 2004.

100 'Reframe Liability Talks with Russia, US Nuclear Vendors Tell Rice', *Nuclear Fuel*, 6 June 2005, at http://construction.ecnext.com/coms2/summary_0249-47259_ITM_platts. For a discussion of possible ways to overcome the liability dispute, see R. Douglas Brubaker and Leonard S. Spector, 'Liability and Western Nonproliferation Assistance to Russia: Time for a Fresh Look?', *Nonproliferation Review*, vol. 10, no. 101, Spring 2003, pp. 1–39.

101 For more on the proposed agreement, see 'Liability Agreement with Russia: A Reversal of US Hard Line', *RANSAC Press Release*, 21 July 2005; Simon Saradzhyan, 'Key Nuclear Dispute Is Resolved', *Moscow Times*, 20 July 2005; 'Sen. Domenici: US, Russia Agree on Liability for Plutonium Disposition', *Inside the Pentagon*, 21 July 2005; and Office of Senator Pete Domenici, 'Long-Awaited US–Russia Plutonium Liability Agreement Is Critical Step in Right Direction', 19 July 2005, at http://domenici.senate.gov/news/printrecord.cfm?id=240897. As of mid-October 2005, the Russian government was still evaluating the text of the proposed agreement,

which had already been cleared by the US National Security Council; see 'Moscow's OK for Cooperative Security Liabilities Deal Expected Soon', *Inside the Pentagon*, 13 October 2005.

102 See for example Task Force of the Secretary of Energy Advisory Board, *A Report Card on the Department of Energy's Nonproliferation Programs with Russia*, January 2001, p. 2; Bunn, Wier and Holdren, *Controlling Nuclear Warheads*, pp. xi, xiv, 38–39, 42, 93, 99–101; Richard A. Clarke et al., *Defeating the Jihadists: A Blueprint for Action* (New York: Century Foundation, 2004), p. 7; and *Agenda for Security: Controlling the Nuclear Threat* (Washington DC: Center for American Progress, February 2005), p. 15.

103 Office of the White House Press Secretary, 'Statement on Nuclear Security Cooperation with Russia', 30 June 2005, at http://www.whitehouse.gov/news/releases/2005/06/20050630-4.html.

104 A Russian study has highlighted the value of encouraging presidential involvement in the bilateral relationship: 'The presidents are more supportive of closer cooperation between the two countries than probably 90 percent of their respective bureaucracies'. See Kuchins et al., *US–Russian Relations*, p. 15. The role of former President Clinton in promoting bilateral threat-reduction initiatives is highlighted in Ashton B. Carter and William J. Perry, *Preventive Defense: A New Security Strategy for America* (Washington DC: Brookings Institution Press, 1999), p. 76. The importance of individual initiative became apparent in 2003, when a months-long freeze in new CTR projects only ended after Senator Richard Lugar personally lobbied Bush to intervene to break the congressional logjam responsible for the hiatus; see Martin Schram, 'Tighten Controls on Russian Arsenals', *Newsday*, 13 October 2004.

105 *Weapons of Mass Destruction: Nonproliferation Programs Need Better Integration* (Washington DC: GAO, January 2005).

A helpful matrix on the 'Bureaucratic Overlap and Diffused Responsibility' afflicting the US agencies involved in threat reduction in Russia appears in Brian Finlay and Andrew Grotto, *The Race To Secure Russia's Loose Nukes: Progress Since 9/11* (Washington DC: Henry L. Stimson Center and Center for American Progress, September 2005), p. 19.

106 Cited in Joe Fiorill, 'New DHS Office To Have Some Authority over Threat Reduction, Export Controls, Says Acting Head', *Global Security Newswire*, 21 April 2005. The reorganisation of the State Department's non-proliferation and arms-control offices will require new mechanisms of interagency cooperation. The restructuring is discussed in Office of the DOS Spokesman, 'Beginning To Transform the State Department To Meet the Challenges of the 21st Century', 29 July 2005, at http://www.state.gov/r/pa/prs/ps/2005/50371.htm; and Wade Boese, 'State Department Announces Reorganization', *Arms Control Today*, vol. 35, no. 8, October 2005, pp. 33.

107 John Mintz and Joby Warrick, 'US Unprepared Despite Progress, Experts Say', *Washington Post*, 8 November 2004.

108 Einhorn and Flournoy (eds), *Protecting Against the Spread of Nuclear, Biological, and Chemical Weapons*, vol. 4, pp. 11, 129. See also the assessment of British BW expert Derek Averre in *Next Generation Threat Reduction: Bioterrorism's Challenges and Solutions*, report of the third meeting of the New Defence Agenda's Bioterrorism Reporting Group (Brussels: Bibliotheque Solvay, 25 January 2005), pp. 51–52.

109 For more on Putin's decision, and an assessment of how to coordinate threat-reduction programmes within the Russian government, see Einhorn and Flournoy (eds), *Protecting Against the Spread of Nuclear, Biological, and Chemical Weapons*, vol. 4, pp. 19–20.

110 Matthew Bouldin, 'Russian Government Restructuring and the Future of WMD Threat Reduction Cooperation: A Preliminary Analysis', *RANSAC Policy Update*, March 2004, at http://www.ransac.org; and Cristina Chuen, 'The 2004 Russian Government Reforms', 13 July 2004, at http://cns.miis.edu/pubs/week/040713.htm.

111 'Russian Government Restructuring and the Future of WMD Threat Reduction Cooperation', *RANSAC Policy Update*, May 2004, at http://www.ransac.org.

112 'New Developments', *Global Partnership Update*, no. 4, May 2004, p. 4, at http://www.sgpproject.org/publications.

113 Kudrik et al., *The Russian Nuclear Industry*, pp. 27–28.

114 Ivan Safronov, 'Sergey Ivanov Otvetit za Poluraspad', *Kommersant*, 10 August 2004.

115 GAO, *Cooperative Threat Reduction: DOD Has Improved*, p. 21.

116 GAO, *Weapons of Mass Destruction: Additional Russian Cooperation Needed to Facilitate US Efforts to Improve Security at Russian Sites*, p. 53.

117 *Ibid.*, p. 40.

118 *RIA Novosti*, 10 September 2004, reprinted in *Yaderniy Kontrol': Informatsiya*, 8–15 September 2004, at http://www.pircenter.org/data/publications/yki32-2004.html.

119 The utility of the Gore–Chernomyrdin Commission is evident in the accounts of James M. Goldgeier and Michael McFaul, *Power and Purpose: US Policy Toward Russia after the Cold War* (Washington DC: Brookings Institution, 2003), pp. 106, 158, 164; and Strobe Talbott, *The Russia Hand: A Memoir of Presidential Diplomacy* (New York: Random House, 2002), pp. 59, 69, 142. The body's formal title was the US–Russian Bi-national Commission on Economic and Technological Cooperation.

120 Michael McFaul, *Re-engaging Russia and Russians: New Agenda for American Foreign Policy* (Washington DC: Center for American Progress, 25 October 2004), p. 5.

121 Gottemoeller, 'Arms Control in a New Era', pp. 50–51.

122 David Smigielski, 'An Overview of the 2002 CT Certification Crisis', Russian–American

Nuclear Security Advisory Council, April 2003, at http://www.ransac.org.

123 'Senate Passes Legislation To Lift CTR Funding Restrictions in Russia', *Inside Missile Defense*, 3 August 2005.

124 For more on the new performance measures for CTR projects see DOD, *Cooperative Threat Reduction Annual Report to Congress: Fiscal Year 2005*, Washington DC, 2004, pp. 8, 16.

125 GAO, *Cooperative Threat Reduction: DOD Has Improved*, Appendix II, pp. 28–29.

126 The Commission on America's National Interests and Russia, *Advancing American Interests and the US–Russian Relationship* (Washington DC: The Nixon Center, September 2003), at http://www.nixoncenter.org/publications/monographs/fr.htm.

Chapter Two

1 State Department, 'Rice Calls United States, Russia Strategic Partners, Not Rivals; US–Russian Relations "Very Warm"', 21 April 2005, at http://www.usembassy.it/file2005_04/alia/a5042105.htm.

2 For a history of US efforts to pursue contacts with the Soviet, Russian and East European militaries, see Marybeth Peterson Ulrich, *Democratizing Communist Militaries: The Cases of the Czech and Russian Armed Forces* (Ann Arbor, MI: University of Michigan Press, 1999), pp. 50–66.

3 Goldgeier and McFaul, *Power and Purpose*, p. 322.

4 DOS Fact Sheet, 'US Assistance to Russia – Fiscal Year 2005', 1 June 2005, at http://www.state.gov/p/eur/rls/fs/48458.htm.

5 Interfax, 'Russian–US Bases in Kyrgyzstan Don't Hinder Each Other – Ivanov', *Central Asia-Caucasus Analyst*, 6 April 2005. Ivanov elaborated by observing: 'The American base in Manas was set up to support the anti-terrorist operation in Afghanistan and the Russian one in Kant to tighten security of the CIS Collective Security Treaty Organization'.

6 See for example the map in *Central Asia in US Strategy and Operational Planning: Where Do We Go From Here?* (Washington DC: Institute for Foreign Policy Analysis, February 2004), p. 66, at http://www.ifpa.org/pdf/S-R-Central-Asia-72dpi.pdf. For more on Russian and US activities in Central Asia, see Kim Murphy, 'Rivalry Brews in Russia's Backyard', *Los Angeles Times*, 4 December 2004; and Richard Weitz, 'Storm Clouds Over Central Asia: Revival of the Islamic Movement of Uzbekistan (IMU)?', *Studies in Conflict and Terrorism*, vol. 27, no. 6, November–December 2004, pp. 465–89.

7 Nathan Hodge, 'Amid Regional Uncertainty, Officials Review Caspian Guard Initiative', *Defense Daily*, 4 August 2005; and Joshua Kucera, 'US Helps Forces, Gains Foothold in Caspian Region', *Jane's Defence Weekly*, 25 May 2005, p. 12.

8 For more on Russian policies regarding military contacts, see Kimberly Marten Zisk, 'Contact Lenses: Transparency and US–Russian Military Ties', *PONARS Policy Memo*, no. 7, October 1997, at http://www.csis.org/ruseura/ponars/policymemos/pm_0007.pdf.

9 *Overcoming Impediments to US–Russian Cooperation*, pp. 93–94.

10 The transparency problem is discussed in Alexei G. Arbatov, 'Military Reform: From Crisis to Stagnation', in Miller and Trenin (eds), *The Russian Military*, pp. 97–98

11 Frank Brown, 'How To Handle Russia?', *Newsweek*, 9 May 2005, p. 40.

12 Felgenhauer, 'New Détente To Die Young'; and 'Russia Cites US Action For War Exercises', *International Herald Tribune*, 11 February 2004.

13 Roger N. McDermott, 'Putin's Military Priorities: The Modernization of the Armed Forces', in Anne C. Aldis and

Roger N. McDermott (eds), *Russian Military Reform: 1992–2002* (London: Frank Cass, 2003), p. 268.

14 Andrew Cottey and Anthony Forster, *Reshaping Defence Diplomacy: New Roles for Military Cooperation and Assistance*, Adelphi Paper 365 (Oxford: Oxford University Press for the IISS, 2004), p. 28.

15 An analysis of how Russia's defence complex coped with the challenges of the 1990s is presented in Julian Cooper, 'The Future Role of the Russian Defence Industry', in Roy Allison and Christopher Bluth (eds), *Security Dilemmas in Russia and Eurasia* (London: Royal Institute of International Affairs, 1998), pp. 94–117; and Richard F. Staar, *The New Military in Russia: Ten Myths That Shape the Image* (Annapolis, MD: Naval Institute Press, 1996), pp. 76–93.

16 Office of Defense Nuclear Nonproliferation, National Nuclear Security Administration, 'Warhead and Fissile Material Transparency (WFMT) Program', at http://www.nnsa. doe.gov/na-20/wfmt.shtml.

17 Scott Peterson, 'Old Weapons, New Terror Worries', *Christian Science Monitor*, 15 April 2004; and Valery E. Yarynich, 'The Ultimate Terrorism', *Washington Post*, 30 April 2004. See also Valery E. Yarynich, *C3: Nuclear Command, Control, and Cooperation* (Washington DC: Center for Defense Information, May 2003); and Bruce G. Blair, 'Hair Trigger Missiles Risk Catastrophic Terrorism', 29 April 2003, at http://www.cdi.org/blair/hair-trigger-dangers.cfm.

18 *Cooperative Threat Reduction Annual Report to Congress: Fiscal Year 2006*, p. 53. The State Department BioIndustry Initiative has similar objectives.

19 Office of the White House Press Secretary, 'Text of the Joint Declaration by President George W. Bush and President Vladimir V. Putin on the New Strategic Relationship Between the United States of America and the Russian Federation', 24 May 2002.

20 *Overcoming Impediments to US–Russian Cooperation*, pp. 9, 44, 63–64, 84–87, 91.

21 Mike Nartker, 'US, Russian Scientists Exploring Collaboration on Floating Nuclear Power Plants', *Global Security Newswire*, 27 August 2004. See also Aleksey Nikol'skiy, 'S 2006 g. Rossiya nachneot stroit' plavuvhie AES', *Vedomosti*, 18 August 2005.

22 State Department, 'United States Initiatives to Prevent Proliferation', 2 May 2005, at http://www.state.gov/t/np/rls/other/45456.htm.

23 Thornton, 'The G8 Global Partnership', pp. 139–141.

24 Mike Nartker, 'United States Supports Expansion of G-8 Nonproliferation Effort, Officials Say', *Global Security Newswire*, 27 April 2004.

25 G8 Senior Group, 'G8 Global Partnership Annual Report', June 2005, p. 8, at http://www.fco.gov.uk/Files/kfile/PostG8_Gleneagles_GPWGAnnualReport2005.pdf.

26 Remarks Prepared for Delivery by Energy Secretary Abraham at the GTRI Partners Conference Opening Keynote Address, 20 September 2004, at http://www.energy.gov/engine/content.do?PUBLIC_ID=16680&BT_CODE=PR_SPEECHES&TT_CODE=PRESSSPEECH.

27 Bunn and Wier, *Securing the Bomb*, p. viii.

28 GAO, *Nuclear Nonproliferation: Sealed Radioactive Sources*; and 'Q & A: Safety and Security of Radioactive Sources', at http://www.iaea.org/NewsCenter/Features/RadSources/radsrc_faq.html.

29 See for example Anna Badkhen, 'Raid in Georgia Triggers "Dirty Bomb" Fears, Police Seize Radioactive Materials', *San Francisco Chronicle*, 17 June 2003.

30 Rob Edwards, 'Risk of Radioactive "Dirty Bomb" Growing', *New Scientist*, 4 June 2004.

31 Previous joint Russian–US HEU removal efforts are reviewed in Cristina Chuen, 'Reducing the Risk of Nuclear Terrorism: Decreasing the Availability of HEU', 6 May 2005, at http://cns.miis.edu/pubs/week/050506.htm; Philipp C. Bleek, 'Global Cleanout: An Emerging Approach

to the Civil Nuclear Material Threat', September 2004, at http://bcsia.ksg. harvard.edu/BCSIA_content/documents/ bleekglobalcleanout.pdf; and 'Working To Eliminate the Nuclear Threat: Past Successes in Removing HEU Stockpiles', 24 March 2004, at http://www.fcnl.org/ issues/arm/nuclear_HEUremoval.htm.

32 Philipp C. Bleek, 'Global Cleanout of Civil Nuclear Material: Toward a Comprehensive, Threat-Driven Response' *SGP Issue Brief*, no. 4, September 2005, pp. 2–3, at http://www.sgpproject.org/publications/SGPIssueBrief/SGP%20Issue%20 Brief%20Bleek.pdf.

33 'United States and Russian Federation Cooperate on Return of Russian-origin Research Reactor Fuel to Russia', *DOE News*, 27 May 2004, at http://www. nnsa.doe.gov/docs/PR_R-04-116_MO UbilateralUSRussiaAgreement(5-04). pdf; and 'Environmental Analysis Completed for Uzbek Spent HEU Return Plan', *Nuclear Fuel*, 18 July 2005, reprinted in 'Strengthening the Global Partnership: Weekly News Roundup', 30 July 2005, at http://www.sgpproject. org/SGP%20News/SGP%20Weekly%20 News%20Roundup%20July%2030%20-%20August%205.pdf.

34 *A More Secure World: Our Shared Responsibility* (New York: United Nations, December 2004), p. 45.

35 UN News Centre, 'Signatories of UN Additional Nuclear Weapons Safeguard Now Number 100', 19 July 2005, at http://www.un.org/apps/news/story. asp?NewsID=5078&Cr=iraq&Cr1=.

36 For a discussion of these and related initiatives, see Perkovich et al., *Universal Compliance*, pp. 37–41.

37 Oleg Volkov, 'ushli na sklad', *Vremya Novostey*, 14 July 2005.

38 *RIA Novosti*, 14 July 2005, reprinted in *Yaderniy Kontrol': Informatsiya*, 13–20 July 2005, at http://www/pircenter.org/data/publications/yki20-2005.html.

39 International Atomic Energy Agency, *Annual Report for 2004* (2005), p. 1, at

http://www.iaea.org/Publications/Reports/Anrep2004/anrep2004_full.pdf.

40 See for example Rumyantsev's comments, *ITAR-TASS*, 7 January 2004, reprinted in *Yaderniy Kontrol': Informatsiya*, 29 December 2003–14 January 2004, at http://www/pircenter.org/data/publications/yki1-2004.html.

41 Kudrik et al., *The Russian Nuclear Industry*, p. 15.

42 Natal'ya Kornelyuk, 'Rasshcheplenie Rinka', *Profil'*, 1 November 2004; and Alena Kornisheva, 'Amerika ne puskaet Rossiyu na rinok OYAT', *Kommersant*, 10 November 2003.

43 The issues raised by such proposals are assessed in Bunn, Wier and Holdren, *Controlling Nuclear Warheads and Materials*, pp. 110–111; and Diaconu and Maloney, 'Russian Commercial Nuclear Initiatives', pp. 107–110. For a discussion of the opposition within Russia, see Julian Evans, 'Alarm Over Radioactive Waste Site', *Moscow Times*, 15 July 2005; and Andrew Osborn, 'Siberia Could Become the World's Atomic Waste Dump, Warn Greens', *The Independent*, 3 May 2005. See also Bellona Foundation, 'Russian Nuclear Storage Facilities Almost Filled Up', 22 June 2005, at http://www. bellona.no/en/international/russia/waste-mngment/38700.html.

44 The expected growth in the worldwide use of civilian nuclear power is discussed in IAEA, *Nuclear Technology Review – Update 2005*, no. GC(49)/INF/3, 11 July 2005, at http://www.iaea.org/About/Policy/GC/GC49/Documents/gc49inf-3.pdf. For an optimistic discussion of Russian firms' involvement in this process see Paul Starobin, 'Springtime for Russia's Nuclear Industry?', *Business Week*, 6 May 2003, p. 36. A more pessimistic assessment can be found in Diaconu and Maloney, 'Russian Commercial Nuclear Initiatives', pp. 97–112.

45 'Russians Push IAEA as Vehicle for Multilateral Fuel Cycle Venture', *Nuclear Fuel*, 18 July 2005, reprinted in

'Strengthening the Global Partnership: Weekly News Roundup', 30 July 2005, at http://www.sgpproject.org/SGP%20News/SGP%20Weekly%20News%20Roundup%20July%2030%20-%20August%205.pdf.

46 Ambassador Linton Brooks, 'Preventing Nuclear Terrorism: Towards an Integrative Approach', remarks delivered to the International Conference on Nuclear Security, London, 16 March 2005, at http://usembassymalaysia.org.my/wf/wf0321_nuclear_prevent.htm.

47 See for example James E. Goodby et al., *Cooperative Threat Reduction for a New Era* (Washington DC: Center for Technology and National Security Policy, National Defense University, September 2004).

48 The barriers to applying CTR-like programmes outside the former Soviet Union are discussed in Sharon Squassoni, *Globalizing Cooperative Threat Reduction: A Survey of Options* (Washington DC: Congressional Research Service, July 2004).

49 Vladimir Ivanov, 'Sergey Ivanov dolozhil prezidentu o svoikh uspekhakh v protivoraketnoy oborone', *Nezavisimoe Voennoe Obozrenie*, 10 December 2004; and Ivan Safronov, 'Rossiya zapustila raketu v nikuda', *Kommersant*, 30 November 2004. Details of the system are available in Robert S. Norris and Hans M. Kristensen, 'Russian Nuclear Forces, 2004', *Bulletin of the Atomic Scientists*, vol. 60, no. 4, July–August 2004, p. 74; and Pavel Podvig (ed.), *Russian Strategic Nuclear Forces* (Cambridge, MA: MIT Press, 2001), pp. 413–418.

50 For a review of Russian–US cooperation and disagreement regarding BMD during the Yeltsin period, see Stephen J. Cimbala, *Russia and Armed Persuasion* (Lanham, MD: Rowman and Littlefield, 2001), pp. 94–96; Goldgeier and McFaul, *Power and Purpose*, pp. 288–293; and Jennifer G. Mathers, *The Russian Nuclear Shield from Stalin to Yeltsin* (London: Macmillan, 2000), pp. 151–174.

51 Cited in Joseph Ferguson, 'De facto Alliance or Temporary Rapprochement?', *Comparative Connections: An E-Journal on East Asian Bilateral Relations*, Autumn 2001, at http://www.csis.org/pacfor/cc/0104Qus_rus.html.

52 Baker and Glasser, *Kremlin Rising*, pp. 134–136.

53 Vladimir Shlapentokh, 'Is the "Greatness Syndrome" Eroding?', *Washington Quarterly*, vol. 25, no. 1, Winter 2002, p. 139.

54 Gottemoeller, 'Nuclear Weapons in Current Russian Policy', p. 205.

55 Pavel Podvig, 'A History of the ABM Treaty in Russia', *PONARS Policy Memo*, no. 109, February 2000, pp. 3–4, at http://www.csis.org/ruseura/ponars/policymemos/pm_0109.pdf.

56 Office of the White House Press Secretary, 'Text of the Joint Declaration by President George W. Bush and President Vladimir V. Putin on the New Strategic Relationship Between the United States of America and the Russian Federation', 24 May 2002.

57 Marc Selinger, *Aerospace Daily & Defense Report*, 13 May 2004, cited in 'United States Looking at Russian Radars, Targets To Help Missile Defense Development', *Global Security Newswire*, 14 May 2004.

58 Rogov et al., *Reducing Nuclear Tensions*, p. 16.

59 See for example the comments of Russian Deputy Foreign Minister Sergey Kislyak in Katerina Labetskaya, 'Sobbitiya v Irake daleko ne uluchshili oshshushshenie bezopasnosti', *Vremya Novostei*, 24 September 2003.

60 Alexander Pikayev, 'US–Russian Missile Defense Cooperation: Limits of the Possible', *PONARS Policy Memo*, no. 315, pp. 2–3, available at http://www.csis.org/ruseura/ponars/policymemos/pm_0315.pdf.

61 Morten Bremer Maerli, 'US–Russian Naval Security Upgrades: Lessons Learned and the Way Ahead', *Naval War College Review*, vol. 56, no. 4, Autumn 2003, p. 27.

62 Vladimir Dvorkin, *The Russian Debate on the Nonproliferation of Weapons of Mass*

Destruction and Delivery Vehicles, BCSIA Discussion Paper 2004-04 (Cambridge, MA: Kennedy School of Government, Harvard University, April 2004), pp. 21–23; and Sebastian Sprenger, 'US, Russian Officials Restart Talks on Joint Data Exchange Center', *Inside the Pentagon*, 31 March 2005.

63 This dialogue is described in Sharon Weinberger, 'Weldon Calls for Foreign Industry, Government To Work with Congress', *Defense Daily*, 14 September 2004. See also Mike Nartker, 'United States Should Maintain Involvement in Missile Defense, US Lawmaker Says', *Global Security Newswire*, 18 June 2004.

64 Jeremy Singer, 'Weldon Presses Missile Defense Cooperation with Russia', *Space News*, 7 March 2005, p. 6.

65 Russian Ministry of Foreign Affairs, 'Press Statement and Answers to Questions, Plesetsk', 18 February 2004, reprinted by The Acronym Institute, Disarmament Documentation, at http://www.acronym.org.uk/docs/0402/doc30.htm.

66 Cited in *ITAR-TASS*, 21 May 2003, reprinted in *Yaderniy Kontrol': Informatsiya*, 15–22 May 2003, at http://www.pircenter.org/data/publications/yki17-2003.html.

67 See for example Ivanov's remarks in *ibid.*, and *ITAR-TASS*, 12 July 2003, reprinted in *Yaderniy Kontrol': Informatsiya*, 12–19 June 2003, at http://www.pircenter.org/data/publications/yki21-2003.html.

68 Cited in 'Tsitata Nomera', in *Yaderniy Kontrol': Informatsiya*, 3–10 August 2005, at http://www.pircenter.org/data/publications/yki23-2005.html.

69 Nicole C. Evans, 'Missile Defense: Winning Minds, Not Hearts', *Bulletin of the Atomic Scientists*, vol. 60, no. 5, September–October 2004, p. 50.

70 'Status Report Requested of RAMOS Program', *CDI Missile Defense Update*, 8 July 2003.

71 'US–Russian Missile Defense Cooperation Should Start With Smaller Projects, Former MDA Chief Says', *Global Security Newswire*, 13 July 2004. Russian attitudes towards RAMOS are examined in Pavel Podvig, 'US–Russian Cooperation in Missile Defense: Is It Really Possible?', *PONARS Policy Memo*, no. 316, November 2003, pp. 2, 4–5, available at http://www.csis.org/ruseura/ponars/policymemos/pm_0316.pdf.

72 S.A. Popov, 'Means and Methods of Overcoming Barriers in Cooperation', in *Strengthening US–Russian Cooperation on Nuclear Nonproliferation*, p. 139.

73 Associated Press, 'Russia Issues Warning on Space-Based Weapons', *Baltimore Sun*, 3 June 2005.

74 See for example Peter Finn, 'Putin: Russia To Deploy Missiles "Unlikely to Exist" Elsewhere', *Washington Post*, 18 November 2004; and Yliya Petrovskaya et al., 'Diplomatiya "Yadernogo nederzhaniya"', *Nizavisimaya Gazeta*, 18 November 2004.

75 See for example Kim Murphy, 'Russia Tests Missile That Could Evade US Defense', *Los Angeles Times*, 19 February 2004.

76 These episodes are recounted in Goldgeier and McFaul, *Power and Purpose*, pp. 158–166, 176–181, 299–304.

77 For a history of Russian–Iranian nuclear cooperation, see Gleb Ivashentsov, 'Rossiya-Iran: Gorizonti partnerstva', *Mezhdunarodnaya Zhizn'*, 22 October 2004; Victor Mizin, 'The Russia–Iran Nuclear Connection and US Policy Options', *Middle East Review of International Affairs*, vol. 8, no. 1, March 2004, at http://meria.idc.ac.il/journal/2004/issue1/jv8n1a7.html; and Vladimir A. Orlov and Alexander Vinnikov, 'The Great Guessing Game: Russia and the Iranian Nuclear Issue', *Washington Quarterly*, vol. 28, no. 2, Spring 2005, pp. 49–66. According to Rose Gottemoeller, Russian officials have shown increasing concern about Iran's nuclear ambitions; see 'The Unexpected Nonproliferation Partner', *Moscow Times*, 16 February 2005. For an opposite view see Ray Takeyh and Nikolas K. Gvosdev, 'Why Rice's Moscow Visit Failed', *Moscow Times*, 20 August 2005.

[78]   Office of the Press Secretary, 'Roundtable Interview of the President by Foreign Print Media', The White House, 5 May 2005. Rice also spoke approvingly of Russia's insistence on the return of its spent fuel from Iran during her visit to Moscow in April 2005.

[79]   Katrin Bennhold, 'Russia Backs Initiative from Europe on Iran', International Herald Tribune, 22 January 2005.

[80]   Cited in Alex Rodriguez, 'Russia: Iran Softens Nuclear Stance', Chicago Tribune, 22 June 2003. See also Simon Saradzhyan and Caroline McGregor, 'Russia Hardens Stance Toward Iran', Moscow Times, 22 September 2003.

[81]   Robert J. Einhorn and Gary Samore, 'Ending Russian Assistance to Iran's Nuclear Bomb', Survival, vol. 44, no. 2, Summer 2000, pp. 61–62.

[82]   Valeria Korchagina, 'Moscow Reaches Out To Tehran', Moscow Times, 29 June 2005.

[83]   Nikolas Gvosdev and Ray Takeyh, 'Cooperating on Iran', ibid., 8 April 2004.

[84]   Rose Gottemoeller, 'A Promising Direction for G8 Leadership', ibid., 8 July 2005.

[85]   State Department, Adherence and Compliance, p. 106.

[86]   Ibid., p. 108.

[87]   Anatoly Medetsky, 'Defense Plant Hit With US Sanctions', Moscow Times, 26 July 2005.

[88]   Lyubov Pronina, 'Russian Military's Combat Potential Unrealized', Defense News, 13 December 2004.

[89]   Cited in Sergey Babkin and Michael Timofeev, ITAR-TASS, 10 February 2005, reprinted in Yaderniy Kontrol': Informatsiya, 11–18 February 2005, at http://www.pir-center/org/data/publications/yki3-2005.html.

[90]   Aleksandr Kraswulin, 'Vizit za okean dlya obmena opitom', Parlamentskaya Gazeta, 31 May 2005. For GAO's assessment, see Defense Trade: Arms Export Control Vulnerabilities and Inefficiencies in the Post-9/11 Security Environment, Washington DC, April 2005.

[91]   'Moscow Looks for New Arms Markets', Russia Reform Monitor, 15 June 2005.

[92]   Vikto Litkovkin, 'Between Scylla and Charybdis: The Russian Defence Industry's Chinese Dilemma', Russia Profile, vol. 2, no. 2, March 2005, p. 34.

[93]   Aleksey Khazbiev, 'Voenno-promishlenniy', Ekspert, 4 October 2004.

[94]   For an assessment of Russian export procedures during the Yeltsin period, see Vladimir A. Orlov, 'Export Controls in Russia: Policies and Practices', Nonproliferation Review, vol. 6, no. 4, Autumn 1999, pp. 139–151; and Michael Beck, 'Russia and Efforts To Establish Export Controls', at http://www.uga.edu/cits/documents/html/nat_eval_russia.htm.

[95]   Details of Russia's WMD-related export controls are discussed in Einhorn and Flournoy (eds), Protecting Against the Spread of Nuclear, Biological, and Chemical Weapons, vol. 4, pp. 91–105.

[96]   Herbert J. Ellison, 'Russian–American Relations', in Stephen K. Wegren (ed.), Russia's Policy Challenges: Security, Stability, and Development (Armonk, NY: M.E. Sharpe, 2003), p. 89.

[97]   Derek Averre, Kenneth N. Luongo and Maurizio Martellini (eds), Advancing Bio Threat Reduction: Findings From an International Conference (Washington DC: Russian–American Nuclear Security Advisory Council, 2004), p. 15.

[98]   'Recent Developments in the NIS: 2004 Updates and Changes in NIS Export Control Systems and Legislation', NIS Export Control Observer, April 2005, at http://cns.miis.edu/nis-excon, p. 3.

[99]   'Ukaz 'O Komissii po eksportnomu kontrolyu Rossiyskoy Federatsii'', 26 April 2005, at http://kremlin.ru/text/docs/2005/04/87128.shtml. See also 'President Putin Expands Functions and Modifies Membership of Russian Export Control Commission', NIS Export Control Observer, April 2005, at http://cns.miis.edu/nis-excon; and Ivan Safronov, 'President Delivers Export Control into

the Good Hands of the Defense Minister', *Kommersant*, 26 April 2005, at http://www. kommersant.com/page.asp?id=573695.

100 Igor Khripunov, 'Nuclear Security: Attitude Check', *Bulletin of the Atomic Scientists*, vol. 61, no 1, January–February 2005, p. 60.

101 Bunn and Wier, *Securing the Bomb 2005*, pp. 20–21, 44, 81–82, 108–113.

102 The Russian government's formal rationale for joining the PSI is explained at the website of the Ministry of Foreign Affairs, at http://www.mid.ru. Previous Russian concerns about the PSI are discussed in Alex Rodriguez, 'Russian Hesitant To Sign Pledge To Inspect Planes, Ships for Trafficking', *Chicago Tribune*, 3 February 2004; and Michael Roston, 'Russia and the Proliferation Security Initiative', Russian–American Nuclear Security Advisory Council, 16 March 2004, at http://www. ransac.org.

103 Carla Anne Robbins, 'Nuclear Nonproliferation Efforts Hit Snag', *Wall Street Journal*, 27 January 2005.

104 Sergei Ivanov, 'The World in the 21st Century: Addressing New Threats and Challenges', 13 January 2005, at http://www.cfr.org/pub7611/richard_ n_haass_mikhail_fridman_sergey_ ivanov/the_world_in_the_21st_ century_addressing_new_threats_and_ challenges.php#.

105 US Central Intelligence Agency, *Acquisition of Technology Relating to Weapons of Mass Destruction and Advanced Conventional Munitions, July 1 through December 31, 2003*, November 2004, available at http:// www.cia.gov/cia/reports/721_reports/ july_dec2003.htm.

106 ITAR-TASS, 29 June 2005, reprinted in *Yaderniy Kontrol': Informatsiya*, 22 June 2005, at http://www.pircenter/org/data/ publications/ykil7-2005.html.

107 Mike Nartker, 'Putin Criticizes Nonproliferation Approaches', *Global Security Newswire*, 5 December 2003.

108 The effects of the law are assessed in Sharon Squassoni and Marcia S. Smith, *The Iran Nonproliferation Act and the International Space Station: Issues and Options* (Washington DC: Congressional Research Service, 2 July 2004).

Chapter Three

1 For comprehensive reviews of Russia's opposition and response to NATO's enlargement and military intervention in Kosovo, see Ronald D. Asmus, *Opening NATO's Door: How the Alliance Remade Itself for a New Era* (New York: Columbia University Press, 2002), esp. pp. 175–211; J. L. Black, *Russia Faces NATO Expansion: Bearing Gifts or Bearing Arms?* (Lanham, MD: Rowman and Littlefield, 2000); and Goldgeier and McFaul, *Power and Purpose*, pp. 183–210, 247–266.

2 Dmitri Trenin and Bobo Lo, *The Landscape of Russian Foreign Policy Decision-Making* (Moscow: Carnegie Moscow Center, 2005), p. 4.

3 V.V. Putin, 'Vistuplenie na zasedanii Soveta Bezopasnosti', 28 January 2005, reprinted in *Yaderniy Kontrol': Informatsiya*, 1 January–3 February 2005, at http://www. pircenter/org/data/publications/ykil- 2005.html. The NRC's early achievements are discussed in Paul Fritch, 'Building Hope on Experience', *NATO Review*, no. 3, Autumn 2003, at http://www.nato.int/ docu/review/2003/issue3/english/art3. html.

4 United States Mission to NATO, 'NATO, Russia Enhance Military Cooperation', 15 April 2004, at http://nato.usmission.gov/ Article.asp?ID=70E14E9A-4D76-4469- 8344-278F2E8D0042.

5 Cited in Sergio Rossi, 'The Time Has Come for Frank Dialogue', *Il Sole 24 Ore*, 14 February 2005. See also Sergei Lavrov, 'Democracy, International Governance, and the Future World Order', *Russia in Global Affairs*, vol. 3, no. 1, January–March

2005, p. 151. Russians' previous enthusiasm for a greater security role for the EU is reviewed in Laszlo Poti, 'Putin's European Policy', in Janusz Bugajski (ed.), *Toward an Understanding of Russia: New European Perspectives* (New York: Council on Foreign Relations, 2002), pp. 138–140.

6   The deterioration in Russian–EU relations is reviewed in Yuri Borko, 'Rethinking Russia–EU Relations', *Russia in Global Affairs*, no. 3, July–September 2004, at http://eng.globalaffairs.ru/numbers/8/591.html; and Dov Lynch, 'Struggling with an Indispensable Partner', in Dov Lynch (ed.), *What Russia Sees*, Chaillot Paper 74 (Paris: Institute for Security Studies of the European Union, January 2005), pp. 115–136.

7   See for example Igor S. Ivanov, *The New Russian Diplomacy* (Washington DC: Brookings Institution, 2002), p. 96.

8   Vladimir Socor, 'Moscow Criticizes EU and OSCE over Kyrgyz Election', *Eurasia Daily Monitor*, 25 March 2005. Russia's changing position towards the OSCE is examined in Dmitry Danilov, 'Russia and European Security', in Lynch (ed.), *What Russia Sees*, pp. 91–95; and Richard Sakwa, *Russian Politics and Society*, third edition (London: Routledge, 2002), pp. 415–416.

9   Frances G. Burwell (ed.), *The New Partnership: Building Russia–West Cooperation on Strategic Challenges* (Washington DC: Atlantic Council of the United States, April 2005), pp. 8, 21.

10   R. Nicholas Burns and Alexander Vershbow, 'Istanbul Summit: Building a Partnership with Russia', *International Herald Tribune*, 26 June 2004.

11   For more information on this events see 'Exercise "Kaliningrad 2004"', *NATO Press Release*, (2004)090, 15 June 2004, at http://www.nato.int/docu/pr/2004/p04-090e.htm.

12   The text of the 'NATO–Russia Action Plan on Terrorism' is at http://www.nato.int/docu/basictxt/b041209a-e.htm.

13   Sergei Ivanov, 'Keynote Address at the International Institute for Strategic Studies

– Russia and NATO: Strategic Partners Responding to Emerging Threats', London, 13 July 2004, at http://www.iiss.org/showdocument.php?docID=402.

14   'Remarks by Defense Minister Sergei Ivanov to Foreign Military Attaches and Diplomats in Moscow General Staff Academy', 10 December 2004, *Federal News Service*.

15   See for example Nikolay Poroskov, 'Laboratoriya antiterrora', *Vremya Novostey*, 2 August 2005.

16   For a summary of existing international programmes in this area, see Andrew J. Grotto, *Defusing the Threat of Radiological Weapons: Integrating Prevention with Detection and Response* (Washington DC: Center for American Progress, July 2005), pp. 5–7. The Evian initiative commits G-8 governments to track radioactive sources and cooperate to recover missing sources; improve export controls; increase physical protection; ensure the safe disposal of spent sources; and offer assistance and technical support to other countries.

17   'Gleneagles Statement on Non-Proliferation', 8 July 2005, para. 5, at http://www.fco.gov.uk/Files/kfile/PostG8_Gleneagles_CounterProliferation.pdf.

18   Sarah E. Mendelson, 'US–Russian Military Relations: Between Friend and Foe', *Washington Quarterly*, vol. 25, no. 1, Winter 2002, p. 169.

19   'Russia To Support NATO's Mediterranean Anti-Terrorist Operation', *NATO Update*, 8 December 2004, at http://www.nato.int/docu/update/2004/12-december/21209.htm.

20   Vladimir Ivanov, 'Rossiya i NATO eshchye nesovmestnie' *Nezavisimoe Voennoe Obozrenie*, 19–25 November 2004.

21   See for example Putin's remarks cited in Caroline McGregor, 'President Speaks to Muslim World', *Moscow Times*, 10 October 2003; and Defence Minister Ivanov's comments in an interview in Ol'ga Vandisheva, 'Ministr oboroni Rossii Sergey Ivanov: Esli nado-nanesem preventivnie udari', *Komsomol'skaya Pravda*, 26 October 2004.

A detailed assessment appears in Andrey Zlobin, 'V Tsentre Vnimaniya', *Vremya Novostey*, 10 September 2004.

22   'NATO Secretary General Signs Memorandum of Understanding on Civil Emergency Planning and Disaster Preparedness with Russia', *NATO Press Release*, (96)44, 20 March 1996, at http://www.nato.int/docu/pr/1996/p96-044e.htm.

23   Barry Adams, 'NATO–Russia Relations: The Evolving Culture of Security Cooperation', *University of Georgia Center for International Trade and Security Monitor*, vol. 11, no. 1, Spring 2005, p. 27.

24   Steven R. Weisman, 'NATO Talks, Accord and Discord for US and Russia', *New York Times*, 22 April 2005. See also 'Russia To Join Partnership Status of Forces Agreement', 21 April 2005, at http://www.nato.int/docu/update/2005/04-april/e0421a.htm; and Kseniya Kaminskaya, *ITAR-TASS*, 21 April 2005, reprinted in *Yaderniy Kontrol': Informatsiya*, 29 April–5 May 2005, at http://www.pircenter/org/data/publications/yki9-2005.html.

25   'When Disaster Strikes! – Public Support for State and Local Government Response', Heritage Foundation Panel, 29 April 2005, audio recording at http://www.heritage.org/Press/Events/ev042905a.cfm.

26   Kudrik et al., *The Russian Nuclear Industry*, p. 100. See also *Russian Nuclear Submarines: US Participation in the Arctic Military Environmental Cooperation Program Needs Better Justification* (Washington DC: September 2004); and *Coordinating Submarine Dismantlement Assistance in Russia* (Monterey, CA: Center for Nonproliferation Studies, September 2004), pp. 9–10.

27   For a discussion of the complex negotiations leading to Russia's involvement in NATO's peacekeeping operation in Bosnia, see Ashton B. Carter and William J. Perry, *Preventive Defense: A New Security Strategy for America* (Washington DC: Brookings Institution Press, 1999), pp. 33–45.

28   SFOR employed a slightly different command structure.

29   In an effort to remind Western governments not to ignore Moscow's continued relevance to European security, Russian leaders redeployed some 200 troops from neighbouring Bosnia to Slatina airport near Pristina shortly before NATO peacekeeping forces planned to enter Kosovo. NATO was not, as required, informed in advance. NATO's Supreme Commander, Gen. Wesley Clark, ordered his local commander to use force if necessary to remove the troops, but Lt-Gen. Sir Michael Jackson demurred. The evidence that Yeltsin himself endorsed the deployment decision is presented in Brian D. Taylor, *Politics and the Russian Army: Civil–Military Relations, 1689–2000* (New York: Cambridge University Press, 2003), pp. 315–16. Yeltsin's own description of the affair appears in his memoirs, *Midnight Diaries* (London: Phoenix Press, 2001), p. 266.

30   Russia's peacekeeping experience in the former Yugoslavia is discussed in Alexander Nikitin, 'Partners in Peacekeeping', *NATO Review*, no. 4, Winter 2004, at http://www.nato.int/docu/review/2004/issue4/english/special.html.

31   Ivan Safronov, 'Moskva formiruet brigadu dlya sovmestnix deystviy s NATO', *Kommersant*, 14 April 2004; and 'Meeting of NATO–Russia Council with Military Representatives', 15 March 2005, *IMS Press Advisory*, 18 March 2005, at http://www.nato.int/ims/news/2005/n050318e.htm.

32   For discussions of earlier proposals to form a joint NATO–Russian peacekeeping unit, see Peter B. Zwack, 'A NATO–Russia Contingency Command', *Parameters: US Army War College Quarterly*, vol. 34, no. 1, Spring 2004, pp. 89–91.

33   Vladimir Kuzar', 'Pered nami odin i te zhe ugrozi' [interview with the Director of NATO's Moscow Information Bureau], *Krasnaya Zvezda*, 2 December 2004.

34   'Meeting of the NATO–Russian Council at the Level of Foreign Ministers', *NATO*

Press Release, 9 December 2004, at http://ww.nato.int/docu/pr/2004/p041209e.htm.

35 *Interfax-AVN*, 16 March 2005.

36 Presentation by General James Jones, Commander, US European Command, at the National Press Club, Washington DC, 23 November 2004, *Federal News Service*.

37 'NATO–Russia Interoperability Courses in Moscow', *NATO Update*, 13 October 2004, at http://www.nato.int/docu/update/2004/10-october/e1011b.htm.

38 Harald Kujat, 'Enhancing Interoperability', *Krasnaya Zvezda*, 26 February 2004, available in English translation at http://www.nato.int/docu/'articles/2004/a40226a.htm.

39 Natal'ya Ryazantseva, 'Otvechaya na vizovi vremeni [interview with Yu. N. Baluevskiy, head of the RF General Sraff]', *Krasnaya Zvezda*, 6 November 2004, reprinted in *Yaderniy Kontrol': Informatsiya*, 3–10 November 2004, at http://www.pircenter.org/data/publications/yki40-2004.html.

40 Andrei Zaitsev, 'Russia's Defense Sector Targets NATO Markets', *Izvestia*, 21 April 2005, p. 10.

41 Donald C. Daniel and Michael I. Yarymovych, 'Russia and NATO: Perspectives on Increased Interaction in Defence Research and Technology', unpublished paper. For a general description of the RTO, see Donald C. Daniel and Leigh C. Caraher, *NATO Defense Science and Technology* (Washington DC: Center for Technology and National Security Policy, National Defense University, March 2003).

42 John Reppert, 'Russia's Threat Perceptions', in Stephen K. Wegren (ed.), *Russia's Policy Challenges: Security, Stability, and Development* (Armonk, NY: M.E. Sharpe, 2003), p. 11.

43 Janusz Bugjski, *Cold Peace: Russia's New Imperialism* (Westport, CT: Praeger, 2004), p. 35.

44 Robert E. Hunter and Sergey M. Rogov, *Engaging Russia as Partner and Participant: The Next Stage of NATO–Russia Relations* (Santa Monica, CA: RAND, 2004), pp. 10–11.

45 Cited in Alexander Konovalov, *ITAR-TASS*, 4 November 2004.

46 See for example *Interfax-AVN*, 10 November 2004, reprinted in *Yaderniy Kontrol': Informatsiya*, 3–10 November 2004, at http://www.pircenter/org/data/publications/yki40-2004.html.

47 Pavel Podvig, 'Putin's Boost-Phase Defense: The Offer That Wasn't', PONARS Policy Memo no. 180, November 2000, at http://www.csis.org/ruseura/ponars/policymemos/pm_0180.pdf.

48 Editorial, 'Russia Rethinking Missile Defense', *Chicago Tribune*, 23 February 2001.

49 Daniel and Yarymovych, 'Russia and NATO', p. 4.

50 'First Ever NATO–Russia Missile Defence Exercise', *NATO Update*, 11 March 2004, http://www.nato.int/docu/update/2004/03-march/e0308a.htm. Russia's participation is described in Aleksey Lyashchenko, 'V rezhime real'nogo vremeni', *Krasnaya Zvezda*, 11 March 2004; and Nikolay Poroskov, 'Rossiysko-evropeyskiy "zontik"', *Vremya Novostey*, 12 March 2004.

51 Aleksandr Shishlo, *RIA Novosti*, 9 March 2005, reprinted in *Yaderniy Kontrol': Informatsiya*, 11–18 March 2005, at http://www.pircenter.org/data/publications/yki6-2005.html; and 'NATO and Russia To Conduct Joint Theatre Missile Defence Exercise', *NATO Press Release*, (2005)034, 9 March 2005, at http://www.nato.int/pr/2005/p05-034e.htm.

52 Katerina Labetskaya, 'Sobbitiya v Irake daleko ne uluchshili oshshushshenie bezopasnosti' [Interview with Sergei Kislyak]', *Vremya Novostei*, 24 September 2003.

53 Aleksei Lyaschenki, 'Prgamaticheskaya Vstrecha', *Krasnaya Zvezda*, 16 March 2005.

54 See for example the interview with Yuri Baluyevsky in *Izvestia*, 29 July 2003.

55 Alla Kassianova, 'Missile Defense Cooperation in the US–European-Russian Triangle', PONARS Policy Memo no. 313, November 2003, at http://www.csis.org/

ruseura/ponars/policymemos/pm_0313. pdf; and Alla Kassyanova, 'Russian–European Cooperation on TMD: Russian Hopes and European Transatlantic Experience', *Nonproliferation Review*, vol. 10, no. 3, Autumn/Winter 2003, pp. 1–13.

56 Cited in Sergio Rossi, 'The Time Has Come for Frank Dialogue', *Il Sole 24 Ore*, 14 February 2005. See also Fedor Luk'yanov, 'SSHA vozvodyat raketniy sabor vokrug Rossii', *Rossiyskaya Gazeta*, 19 January 2005.

57 Kassyanova, 'Russian–European Cooperation on TMD'.

58 'Launch of NATO's Active Layered Theatre Ballistic Missile Defence (ALTBMD) Programme', *NATO Press Release*, (2005)036, 16 March 2005, at http://www.nato.int/docu/pr/2005/p05-036e.htm.

59 Ivanov allegedly raised the question with US Defense Secretary Donald Rumsfeld when Ivanov visited Washington in January 2005; see Guy Dinmore, Demetri Sevastopulo and Hubert Wetzel, 'Russia Eyes Withdrawal from Key Treaty To Cut Missiles', *Financial Times*, 9 March 2005; and Artur Blinov, 'Moskva i Vashington popitalis' viyti iz soglasheniya, poloshivshego konets raketomu krizisu v Evrope', *Nezavisimaya gazeta*, 11 March 2005.

60 For a description of the PNI see Kurt Guthe, *The Nuclear Posture Review: How Is the 'New Triad' New?* (Washington DC: Center for Strategic and Budgetary Assessments, 2002), p. 23.

61 See for example Robin Cook and Robert McNamara, 'Is it Time To Dismantle the Cold War's Nuclear Legacy', *Financial Times*, 23 June 2005; and Daryl G. Kimball, 'Small, Portable, Deadly, and Absurd: Tactical Nuclear Weapons', *International Herald Tribune*, 3 May 2005.

62 For example, Section 3621(a) of the FY 2004 National Defense Authorization Act (PL 108-136) says that 'the United States should, to the extent the President considers prudent, seek to work with the Russian Federation to develop a comprehensive inventory of Russian tactical nuclear weapons'.

63 Hearings of the Senate Armed Services Committee on 'The Strategic Offensive Reductions Treaty', 25 July 2002, *Federal News Service*.

64 Cited in Wade Boese, 'Deeper Nuclear Cuts Unlikely for Now', *Arms Control Today*, vol. 35, no. 6, July–August 2005, p. 36.

65 See for example the remarks of Colonel General Leonid Ivashov cited in 'Psledniy sekret', *Nezavisimoye Voyennoye Obozreniye*, 10–16 June 2005. David S. Yost has identified nine specific functions for tactical nuclear weapons in his survey of Russian commentaries on their possible uses; see his 'Russia's Non-Strategic Nuclear Forces', *International Affairs*, vol. 77, no. 33, 2001, pp. 531–551, esp. pp. 534–537. For a discussion of the size and character of Russia's tactical nuclear-weapons arsenal, as well as the problems involved in defining a 'tactical nuclear weapon', see Gunnar Arbman and Charles Thornton, *Russia's Tactical Nuclear Weapons, Part 1: Background and Policy Issues* (Stockholm: Swedish Defence Research Agency, November 2003).

66 Cited in Robert S. Norris and Hans M. Kristensen, 'NRDC: Nuclear Notebook: US Nuclear Forces, 2004', *Bulletin of the Atomic Scientists*, vol. 60, no. 3, May–June 2004, p. 69.

67 Cited in Associated Press, 'Russia Issues Warning on Space-Based Weapons', *Baltimore Sun*, 3 June 2005. On the number of air-deliverable US nuclear bombs currently deployed in Europe, see the report by the Natural Resources Defense Council, *US Nuclear Weapons in Europe: A Review of Post-Cold War Policy, Force Levels, and War Planning*, February 2005, at http://www.nrdc.org/nuclear/euro/contents/asp.

68 Cited in Boese, 'Deeper Nuclear Cuts Unlikely', p. 36.

69 See for example the remarks of Vladimir Verkhovtsev, deputy chief of the MOD 12th Main Directorate, in 'US Senator Seeks

Access to Russia's Nuclear Weapons', *Izvestia*, 1 June 2005.

70 Current NSC policy guidance prohibits US assistance to operational warhead storage sites because it could enhance Russia's military capacities; see GAO,

Nonproliferation Programs Need Better Integration, p. 20.

bs Nikolay Poroskov, 'Vzrivnie Eksperimenti', *Vremya Novostey*, 10 August 2004; and Ivan Safronov, 'Sergey Ivanov Otvetit za Poluraspad', *Kommersant*, 10 August 2004.

Conclusion

1 James M. Goldgeier, 'Dissuasion in America's Russia Policy', *Strategic Insights*, vol. 3, no. 10, October 2004, at http://www.ccc.nps.navy.mil/si/204/oct/goldgeierOct04.asp.

2 Sergei Blagov, 'Iranian Nuclear Team in Moscow, Seeking New Partnerships', *Eurasia Daily Monitor*, 12 July 2005.

3 William Hoehn, 'Update on Congressional Activity, pp. 25, 26 at http://www.ransac.org/Publications/ Congress%20and%20Budget/Federal%20 Budget%20and%20Congressional%20Up dates/index.asp.

4 United Press International, 'Russia Poll Finds Most Prefer US Ties', *Washington Times*, 13 September 2005. See also the July 2005 poll on Russian–US relations summarised in *RIA Novosti*, 15 September 2005, at http://www.russiaprofile.org/cdi/ article.wbp?article-id=8C8FFD0E-B38A-4795-BD7F-2E7DF8B11C(F.